D0204720

BIOTECHNOLOGY IN AGRICULTURE SERIES

General Editor: Gabrielle J. Persley, Biotechnology Adviser, Environmentally Sustainable Development, The World Bank, Washington DC, USA.

For a number of years, biotechnology has held out the prospect for major advances in agricultural production, but only recently have the results of this new revolution started to reach application in the field. The potential for further rapid developments is, however, immense.

The aim of this book series is to review advances and current knowledge in key areas of biotechnology as applied to crop and animal production, forestry and food science. Some titles focus on individual crop species, others on specific goals such as plant protection or animal health, with yet others addressing particular methodologies such as tissue culture, transformation or immunoassay. In some cases, relevant molecular and cell biology and genetics are also covered. Issues of relevance to both industrialized and developing countries are addressed and social, economic and legal implications are also considered. Most titles are written for research workers in the biological sciences and agriculture, but some are also useful as textbooks for senior-level students in these disciplines.

Editorial Advisory Board:

P.J. Brumby, formerly of the World Bank, Washington DC, USA.

E.P. Cunningham, Trinity College, University of Dublin, Ireland.

P. Day, Rutgers University, New Jersey, USA.

J.H. Dodds, International Center for Agricultural Research in the Dry Areas, Syria.

J.J. Doyle, formerly of the International Laboratory for Research on Animal Diseases, Nairobi, Kenya.

S.L. Krugman, United States Department of Agriculture, Forest Service.

W.J. Peacock, CSIRO, Division of Plant Industry, Australia.

BIOTECHNOLOGY IN AGRICULTURE SERIES

Titles Available:

Intellectual Property Rights in Agricultural Biotechnology

Edited by

F.H. Erbisch
Director, Office of Intellectual Property
Michigan State University
USA

and

K.M. Maredia
Technology Transfer Coordinator and Associate Professor
Institute of International Agriculture
Michigan State University
USA

CAB INTERNATIONAL

CAB INTERNATIONAL
Wallingford
Oxon OX10 8DE
UK

Tel: +44 (0)1491 832111
Fax: +44 (0)1491 833508
E-mail: cabi@cabi.org

CAB INTERNATIONAL
198 Madison Avenue
New York, NY 10016-4314
USA

Tel: +1 212 726 6490
Fax: +1 212 686 7993
E-mail: cabi-nao@cabi.org

A catalogue record for this book is available from the British Library, London,
UK

Library of Congress Cataloging-in-Publication Data
Intellectual property rights in agricultural biotechnology / edited by
 F.H. Erbisch and K.M. Maredia.
     p.   cm. -- (Biotechnology in agriculture series; 20)
   Includes index.
   ISBN 0-85199-232-3 (alk. paper)
   1. Agricultural biotechnology--Patents. 2. Agricultural
biotechnology--Law and legislation.   I. Erbisch, Frederic H.
II. Maredia, Karim M.   III. Series.
K1519.B54I58   1998
346.04'8--dc21   97-32257
                                       CIP

ISBN 0 85199 232 3

Typeset in 10/12pt Melior by Columns Design Ltd, Reading
Printed and bound in the UK by Biddles Ltd, Guildford and King's Lynn

Contents

Contributors

Elinor Arteaga-Marcano, CONICIT, Edf. Malploca, Av. Principal de los Cortijos, Caracas, Venezuela.

Andrei A. Baev, Mayer, Brown and Platt, 25th Floor, 350 South Grand Avenue, Los Angeles, CA 90071-1503, USA.

John H. Barton, George E. Osborne Professor of Law, Stanford University, Crown Quadrangle, Stanford, CA 94305, USA.

Michael Blakeney, Asia Pacific Intellectual Property Law Institute, School of Law, Murdoch University, Murdoch, WA 6150, Australia.

Rosario Castañón, Center for Technological Innovation, National University of Mexico, Ciudad Universitaria, PO Box 22510, Mexico.

R. Stephen Crespi, Patent Consultant, 16 Kenlegh, Bognor Regis PO21 3TS, UK.

Atef El-Azab, Academy of Scientific Research and Technology, 18 Midan El Misaha, Dokki, Giza, Egypt 12311.

Frederic H. Erbisch, Office of Intellectual Property, 238 Administration Building, Michigan State University, East Lansing, MI 48824-1046, USA.

Andrew J. Fischer, Office of Intellectual Property, 238 Administration Building, Michigan State University, East Lansing, MI 48824-1046, USA.

Prabuddha Ganguli, Corporate Planning, Hindustan Lever Ltd, Hindustan Lever House, 165/166 Backbay Reclamation, Mumbai 400 020, India.

Walter R. Jaffe, CONICIT, Edf. Malploca, Av. Principal de los Cortijos, Caracas, Venezuela.

Atushi Komamine, Faculty of Science, Nihon Women's University, 2-8-1 Mejiro-dai, Bunkyo-ku, Tokyo 112, Japan.

Tan Loke-Khoon, Baker & McKenzie, 14th Floor, Hutchinson House, 10 Harcourt Road, Hong Kong.

Karim M. Maredia, Institute of International Agriculture, 416 Plant and Soil Science Building, Michigan State University, East Lansing, MI 48824-1325, USA.

Yoshihiko Nishizawa, Sumitomo Chemical Industry Co. Ltd, 818-1 Hourensan Soenishi-cho, Nara-city, 630 Nara, Japan.

Silvia Salazar, PO Box 8-5750-1000, San José, Costa Rica.

José Luis Solleiro, Center for Technological Innovation, National University of Mexico, Ciudad Universitaria, PO Box 22510, Mexico.

Carlos Velasquez, 524 S Chestnut Apt N20, Lansing, MI 48933, USA.

Kazuo N. Watanabe, Department of Biotechnological Science, Institute of Bioscience and Technology, Kinki University, 930 Nishi-Mitani, Naka-Gunn, Uchita, Wakayama 649-64, Japan.

Rosemary A. Wolson, Science and Technology Policy Research Centre, DPRU, Hiddingh Campus, University of Cape Town, Private Bag Rondebosch 7700, South Africa.

Foreword

The successful application of biotechnology tools has had and is having dramatic effects in some areas of agriculture. These effects are being felt throughout the world in academic, government and industrial communities. The result is the rapid development of a multi-million dollar industry. This work has been going on for more than two decades though recent advances make it appear as though all the development occurred 'overnight'.

The general subject of intellectual property is not new to the global community. For example, the constitution of the United States written over 200 years ago contains a provision through which an inventor is able to protect his/her discovery for a limited period of time. The general subject matter of intellectual property is normally subdivided into patents, trademarks and copyright law.

Of great recent importance was a decision by the US Supreme Court allowing patenting of living organisms. This landmark decision opened the doors to agriculture and medical biotechnology research companies being able to patent their processes and products.

The application of intellectual property law to biotechnology has changed the way that scientists exchange materials and ideas and continues to have an increasing impact on the way that scientific research is performed. This impact, coupled with global agricultural research, has led to substantial need for education and training of scientists worldwide in the basic fundamentals of intellectual property.

This handbook should serve as a useful primer for scientists, administrators and policy makers who wish to know the basics of intellectual property rights. Furthermore, it is hoped this handbook will be valuable to these individuals throughout the world, especially in the developing world.

The handbook grew out of a series of workshops and short term training courses which have been offered by Michigan State University (MSU) in the application of intellectual property rights law to agricultural research. MSU looks forward to continuing its interest in this important area as both the law and the science move forward into the millennium.

Peter M. McPherson
President,
Michigan State University

Preface

In 1991 Michigan State University entered into a multi-year cooperative agreement with the US Agency for International Development. The objective of the award was, with the assistance of other universities and private industry, to develop research relationships with emerging countries to train their scientists effectively to utilize biotechnology in enhancing plant agricultural products. The project was called the Agricultural Biotechnology for Sustainable Productivity (ABSP) program.

The cooperative research effort proved to be beneficial for all parties. However, it was realized that two non-research policy areas needed to be addressed before the full benefit of the research program could be gained. These policy areas were intellectual property rights and biosafety. Workshops and training programs for both areas were developed by ABSP. Intellectual property workshops were held in the USA, Egypt, Indonesia and Morocco. The responses to these workshops were very positive and, as a result, Michigan State University, with the assistance of ABSP, designed and conducted two intellectual property internship programs at its East Lansing, Michigan, campus. Over 500 individuals, including scientists, attorneys, government officials and other agricultural personnel, from more than 15 emerging countries, participated in these workshops and internship programs.

Participants at these workshops and internship programs often asked about the availability of printed material or a handbook containing the basic materials taught in the program. They wanted to share this with others who they believed would benefit from this material. While handouts were provided, they did not satisfy these requests. Nothing satisfactory was found in published literature, so it was decided to draft a book which would meet the needs of the participants of the workshops and internships. The result is this book. It contains basic information about intellectual property, including its protection and marketing. Special efforts were taken to make the book definitive, yet to minimize the legal jargon which is found in so many published works on intellectual property.

Finally, individuals from around the world were asked to provide a summary of intellectual property management in their country or region. The material provided by these authors illustrates the developmental stage of intellectual property programs, laws and legislation in their geographic regions. It is hoped this materials can provide direction, and perhaps assistance, to those countries developing their own intellectual property programs.

Acknowledgements

The editors would first like to thank CAB INTERNATIONAL for publishing this handbook. Without their patience and expertise, this publication would not have been possible. The Office of Intellectual Property, the Agricultural Biotechnology for Sustainable Productivity project, the Institute of International Agriculture at Michigan State University and their support staff were also very helpful. Special thanks go to all the international contributors for providing insight to the current status of intellectual property rights in their countries or regions.

It is difficult to describe the monumental efforts required to develop this handbook. The assistance of Dr John Dodds from Michigan State University and Professor John Barton from Stanford Law School was instrumental in the development of the initial handbook outline and identifying some of the international contributors.

We also appreciate the assistance of Mr Wade Mitchell, a former Thomas M. Cooley Law School intern, for working with us in the handbook's beginning. The task was then assigned to Mr Andrew Fischer, also a Thomas M. Cooley Law School intern. With his tireless efforts, Mr Fischer worked with us finding additional international authors, coordinating correspondence, and compiling the final product.

Frederic H. Erbisch
Karim M. Maredia
East Lansing, Michigan

List of Acronyms and Abbreviations

ABSP	Agricultural Biotechnology for Sustainable Productivity
AGERI	Agricultural Genetic Engineering Research Institute (Egypt)
AIC	Administration for Industry and Commerce (People's Republic of China)
APEC	Asia Pacific Economic Cooperation
ARC	Agricultural Research Council (South Africa)
ARIPO	African Regional Industrial Property Organization
ASTA	American Seed Trade Association
AUTM	Association of University Technology Managers
BIO	Biotechnology Industry Organization
BIRPI	United International Bureaux for the Protection of Intellectual Property
BST	bovine somatotropin
CBD	Convention on Biological Diversity
CCPA	Court of Customs and Patent Appeals (USA)
CGIAR	Consultative Group on International Agricultural Research
CITES	Convention on International Trade in Endangered Species
CRF	Code of Federal Regulations (USA)
CRIFC	Central Research Institute for Food Crops (Indonesia)
EC	European Community
EPA	Environmental Protection Agency (USA)
EPC	European Patent Convention
EPO	European Patent Office
ETF	enforcement task force
EU	European Union
FAO	Food and Agriculture Organization (UN)
FDA	Food and Drug Administration (USA)
FIE	Foreign Investment Enterprise (People's Republic of China)

GATT	General Agreement on Tariffs and Trade
GMO	genetically modified organism
INPADOC	International Patent Documentation Centre
INPI	National Institute of Industrial Property (Brazil)
IP	intellectual property
IPR	intellectual property rights
M&A	merger and acquisition
MCI	Ministry of Chemical Industry (People's Republic of China)
MOFTEC	Ministry of Foreign Trade and Economic Cooperation (People's Republic of China)
MSU	Michigan State University
MTA	material transfer agreement
NAFTA	North American Free Trade Agreement
NARI	National Agricultural Research Institute
NARS	National Agricultural Research Systems
NCA	National Copyright Administration (People's Republic of China)
NGO	non-governmental organization
NIC	newly industrializing country
NIEO	new international economic order
NTB	non-tariff barriers
OAPI	Organisation Africaine de la Propriété Intellectuelle
OECD	Organization for Economic Cooperation and Development
OIP	Office of Intellectual Property
OTA	Office of Technology Assessment (USA)
PBR	plant breeders' rights
PCT	Patent Cooperation Treaty
PPA	Plant Patent Act (USA)
PRC	People's Republic of China
PTO	Patent and Trademark Office (USA)
PVP	plant variety protection
PVPA	Plant Variety Protection Act (USA)
R&D	research and development
RIFCB	Research Institute for Food Crops Biotechnology in Indonesia
SAGENE	South African Committee for Genetic Experimentation
SAIC	State Administration for Industry and Commerce (People's Republic of China)
SANSOR	South African National Seed Organization
SPC	State Planning Commission (People's Republic of China)
TNC	transnational corporation
TPD	Transvaal Provincial Division
TRIPs	Trade-related Aspects of Intellectual Property Rights (GATT)
TTO	technology transfer office
UN	United Nations
UNCED	United Nations Conference on Environment and Development
UNCTAD	United Nations Conference on Trade and Development
UNCTC	United Nations Centre for Transnational Corporations

UNDP	United Nations Development Program
UNEP	United Nations Environment Program
UNESCO	United Nations Educational, Scientific and Cultural Organization
UNIDO	United Nations Industrial Development Organization
UNU-INTECH	United Nations University Institute for New Technology
UPOV	Union Internationale pour la Protection des Obtentions Végétales
USAID	United States Agency for International Development
USDA	United States Department of Agriculture
USTR	United States Trade Representative
WHO	World Health Organization
WIPO	World Intellectual Property Organization (UN)
WTO	World Trade Organization

Issues and Principles

Introduction to Intellectual Properties

Frederic H. Erbisch[1] and Carlos Velazquez[2]

[1]*Office of Intellectual Property, 238 Administration Building, Michigan State University, East Lansing, MI 48824-1046, USA;* [2]*524 S Chestnut Apt N20, Lansing, MI 48933, USA*

INTRODUCTION

Learning about intellectual property rights (IPR) takes considerable study and continual review of legal literature. However, certain terminology and concepts used in this area are easily learned and assimilated. This basic knowledge is very important in identifying and managing intellectual property (IP). This chapter aims to present basic facts and concepts to enable the scientist, the administrator, the governmental official and the non-intellectual property attorney to recognize and then handle IPs appropriately.

Without this basic knowledge one could lose valuable IPs. For example, when F.H. Erbisch was a researcher at a small university, he was unaware that his research had resulted in an invention. Approximately eight years later, another university annouced a major patented invention which was the same as his invention! The patented invention has been very successful, earned the university and researcher millions of dollars and saved many lives. The author's university had been negligent in educating its researchers and administrative staff in the basics of IP law. This invention may never have been 'lost' to the author had he been instructed properly. Since that time, the author has learned about IP law and, in his administrative roles, has endeavored to educate researchers and administrators so they can properly handle their creations.

While recognition and proper handling of IPs are important, it is also necessary to know when to use those fully trained in IP law. Only through the use of these highly trained individuals will the scientists' discoveries and creations be fully protected. These fully protected ideas can then provide recognition and rewards to the originating organization and the inventor.

In this chapter the concept of IPs will be addressed first. Following this will be a basic discussion of the various means of protecting IPs. All of this will be presented on the basis of how it is done in the USA. The basic premise for each of these means of protection, while perhaps different from those in the USA, is quite similar to those in many other countries.

WHAT ARE INTELLECTUAL PROPERTIES?

In contrast with real property (land) or physical property, which one can see, feel and use, IP is intangible. IPs are ideas and thoughts, or products of the mind. As long as these ideas or thoughts are not expressed in a tangible form, they remain protected and cannot be used by others.

With any type of property there are property rights. When IPs are expressed in a tangible form they can be protected. IPRs have been created to protect the right of individuals to enjoy their creations and discoveries. IPRs can be traced back to the fourteenth century when European monarchs granted proprietary rights to writers for their literary works.

Usually IPR are protected by one of three legal theories: copyrights, trademarks or patents. These theories demonstrate that IPR are private rights. They have been created to ensure protection against unfair trade practice. Owners of IP are granted protection by a state and/or country, under varying conditions and periods of time. This protection includes the right to: (i) defend their rights to the property they created; (ii) prevent others from taking advantage of their ingenuity; (iii) encourage their continuing innovativeness and creativity; and (iv) assure the world a flow of useful, informative and intellectual works.

With the growing recognition of IPR, the importance of worldwide forums on IPs is realized. Worldwide companies, universities and industries want to protect their IPR internationally. In order to reach this goal, countries have signed numerous agreements and treaties and developed orgainizations to oversee their applications. These agreements and treaties include the General Agreement on Tariffs and Trade (GATT), the World Intellectual Property Organization (WIPO), and the Trade-related Aspects of Intellectual Property Rights (TRIPs) treaty.

The desire to promote effective protection of IPR in the international trade forum has grown immensely. All of the previously mentioned agreements have been created to promote a balanced international trading field and to prevent the international trade of counterfeit goods. Another important reason to justify these agreements and their enforcement is the protection of IPR in underdeveloped countries and to enable them to create a sound and viable technological base, allowing participation in international trade. Every country has its own IP laws dealing with patents, trademarks and copyrights. Careful review of these laws should be taken to ensure protection under appropriate jurisdiction. In the USA, protection by patent, trademark and/or copyrights should be utilized to ensure complete protection of every inventor, creator or discoverer's IPR.

WHAT IS COPYRIGHT?

Copyright is often thought of as a special territory for artists, composers and writers. Copyright was created by US Congress in 1807 to provide protection to composers, writers, authors and artists to protect their original works, derivative works and work for hire. Original works are defined as any work showing originality with at least a trace of creative input. Works are not copyrightable if they are mere ideas, transient sounds or gestures; they must be in a tangible form, either visually or audibly, creating the representation of the original work. Derivative works are based on other previously copyrighted work. Derivative works are formed by collecting and assembling pre-existing materials or data, which are then selected, coordinated and/or arranged in such a way that the result as a whole constitutes an original work of authorship. Work for hire deals with ownership right of a work created by a person hired or paid by the legal owner of the work. It is assumed that the employer generally owns the employee's creations (the employee cannot be an independent contractor).

The main consideration in a copyright situation is the concept of originality. The right to protection arises from the original effort and labor of the creator in seeking, arranging and/or listing the content in a new, original form not found in the public domain. Some copyrightable works are in the field of literature, music, drama, choreography, photography, sculpture, audiovisual, sound recordings and even computer program source codes, which are usually registered as literary works. The copyright protection these works receive includes the right to prevent others from reproducing the work, preparing derivative works, distributing copies by sale, rental, lease or lending, and/or performing or displaying the work publicly without the creator's permission. In the USA, copyright law is primarily federal law.

In order to simplify the copyright process in the USA and to make copyright law more accessible, the US Congress, in 1989, passed a resolution which allowed a copyrightable work to be copyrighted once the work was created. Simply putting '© John Doe, 1997' gives notice to others that the material is copyrighted, but the notice is not absolutely mandatory. This makes the copyright the least expensive way to protect IP and allows the creator sole rights in the USA for approximately 75 years. The time period for protection varies; rights for an individual creator are the length of the creator's life plus 50 years during which the creator's estate will have control of any rights on the creator's work. Joint works have a lifetime of the surviving author plus 50 years. Work made for hire protects for 100 years from the creation of the work. The creator must be very careful not to indicate either by express words, actions or by implication, that he/she does not intend to claim a copyright in the work or that he/she has abandoned the work, because the work would then become public and the creator would have no control over it.

With a few exceptions, registration is necessary before a copyright lawsuit can be initiated. The US Copyright Office has required forms which depend upon the work being registered. In addition to the forms, the creator must submit two copies of the required materials and a minimal fee to the Copyright Office. If the application is accepted, the creator will be given a registration number to include with the material when making the work available to others. Registration is a recognition that the creator is fully aware of the value of copyrighting. Upon registration the creator should include a notice such as the phrase, 'All rights reserved', or 'Not to be reproduced without the express permission of the author' with the work. Also, one of the following three markings must be included on the work to identify copyrights in the USA: '©', 'Copy' or the word 'Copyright', followed by the name of the creator and the year of creation. Examples of these markings are '© John Doe 1997', 'Copy John Doe 1997' and 'Copyright John Doe 1997'.

Copyright protection generally is only valid in the country that grants the protection. Many countries respect the copyright of other countries, but if one wants protection in a certain country it is best to apply for a copyright in that country. The agriculture industry uses copyright protection regularly. Directions on use of a product and descriptions of products are just two examples of copyright use.

WHAT IS A TRADEMARK?

A trademark is a symbol that helps to distinguish one product or company from another. Symbols help the consumer identify products and/or a company and include designs, shapes, numbers, slogans, smells, sounds or anything that helps the consumer to identify the products and/or companies.

The trademark is part of the US Unfair Competition Law, which addresses problems like false advertizing and misappropriation. A trademark is different from a tradename: a tradename is used to identify the business entity, whereas the trademark is used to differentiate between a company's product and all other products; in some circumstances it helps to identify the company name. Some examples are the slogan 'Just Do It', which identifies the Nike Company, or the shape of the Coca Cola bottle, which identifies the Coca Cola Company.

The US Supreme Court stated that the primary purpose of trademarks is 'to identify the origin or ownership of the article to which it is affixed', but experts insist that the primary purpose of a trademark is to guarantee that a company's investment in research and development, marketing and the reputation a company has spent years creating in the eye of the consumer is not stolen by a competitor. Some companies have spent millions of dollars in creating an image that is instantly recognized in today's market. This recognition alone is worth millions of dollars. Another benefit of the trademark is that it maintains quality control in products. When a consumer purchases a product identified because of the trademark, the consumer expects a quality product. Companies maintain a high level of quality to remain competitive.

Today's market is so large that it has become very difficult to come up with a trademark which is successfully registered. The Patent and Trademark Office (PTO) has created a registration process to ensure the protection of a company trademark. This process requires a prescreening in trademark directories and catalogues. A professional search is recommended by using one of two computerized databases: TRADE-MARKS CAN and/or COMPUMARK. Once the registration process has begun and a company wants to introduce a product into the market, they are required to use the proper trademark grammar. This grammar communicates to all competitors that the company trademark is registered or is in the registration process. Trademark grammar is to be placed on the product itself or on any advertizing. There are two choices of trademark grammar: Trademark^R or Trademark*, followed by a footnote indicating federal registration. This grammar is essential if infringement occurs.

Trademark law, unlike patent or copyright law, confers a perpetual right. So long as the trademark continues to identify a single source, anyone who uses a very similar mark may be liable for trademark infringement. The perpetual right of trademarks depends on the use. The basic idea of 'use it or lose it' is essential to preserving trademark rights. A company cannot register a trademark and then not use it. The product for which the trademark was registered must be being used commercially or the trademark rights will cease to exist.

Trademark rights are so important that multinational companies spend fortunes to maintain their respective trademarks around the world. Every country has different trademark laws. However, there are agreements to ensure that a company's trademark in one country is protected in another country. The North American Free Trade Agreement

(NAFTA) preserves registration of marks under the trademark law of the given country, but ensures that each member country (Canada, Mexico and the USA) provides uniformity in its trade law. There have been cases when 'pirates' register large US companies' trademarks in their countries, wait until that company markets the product in their country and then charge the companies large amounts of money for the use of their own trademarks.

The PTO receives an average of 100,000 trademark applications each year, and this number is growing rapidly. The PTO takes an average of 12 months to review and grant the trademark. In our consumer-orientated market, the value of a trademark is rising, and with global markets opening every day, the value of an effective trademark will continue to rise.

WHAT IS A PATENT?

A patent is an exclusive right given to an inventor to exclude all others from making, using and/or selling the invention. The right the inventor has depends on which country issued the patent. For an invention to be protected in the USA, an inventor must file a patent application with the US PTO within one year after having it disclosed. Once issued, a patent gives the inventor the legal right to create a monopoly by excluding others from creating, producing or selling the invention. This right to exclude others from the invention is limited to a period of 20 years from the date of filing the patent application.

The purpose of a patent is to promote the progress of science and useful arts. Patent law promotes this progress by giving the inventor the right of exclusion. In exchange for this right to exclude others, the inventor must disclose all details describing the invention, so that when the 20 year patent right expires, the public may have the opportunity to develop and profit from the use of the invention. Specifically, the inventor must disclose the best mode of the invention.

There are three types of patents: (i) plant patent; (ii) design patent; and (iii) utility patent or 'regular patent'.

Plant patents are granted for newly discovered asexually propagated plants. Like a utility patent, the plant patent provides 20 years' protection. Unlike a utility patent, which protects functional characteristics, the design patent protects ornamental characteristics. The lifespan of a design patent is only 14 years. The design patent prevents a competitor from copying a unique design and profiting from it. Examples of companies that deal with design patents are toy, souvenir and industrial manufacturers. The utility patent constitutes the largest portion of all patents issued. It is most commonly used by companies and universities to protect the results of their research and development, and has a lifespan of 20 years from the time of filing with the US PTO. The US Patent Act provides that 'whoever invents or discovers any new and useful process, machine, manufacture, or composition of matter may obtain a patent'. These terms are defined as follows:

1. 'Process' refers to the method used to produce the invention itself.
2. 'Machine' refers to a machine that produces a product of some sort.
3. 'Manufacture' refers to the actual invention itself, whatever the invention is.
4. 'Composition of matter' refers to the composition or formula of an invention that requires it for its existence.

For the US PTO to issue a utility patent, the inventor must establish that the invention is novel (new), is non-obvious to one skilled in the field and has a utility (useful). The novelty requirement refers to the prior existence of an invention. If an invention is identical to an already patented invention, the 'novelty' requirement is not met, so a patent cannot be issued. The 'useful' requirement refers to the practical use of the invention. If the invention provides a product that is required or needed in some manner, then it meets this requirement. If this requirement is not met, a patent will not be issued. The 'non-obvious' requirement refers to the level of difficulty required to invent or discover the technology. If the invention is so obvious that anyone having an ordinary skill would have thought of it, then it probably does not meet this requirement. The main point to consider when assessing this is the situation at the time of the discovery or invention. What might appear obvious once the invention is presented for patent may not have been so obvious before the invention or discovery. If an invention fulfils all three elements, then the invention fulfils the US PTO requirements necessary for patenting.

Invention improvements can also be protected. Even though the improvements may not be patentable on most occasions, as they are not considered novel because of the parent invention, some improvements are so innovative and useful they become inventions to themselves. Improvements that are just too obvious or are not worth patenting because of a number of limitations are considered as 'know-how' of the original invention.

Most biotechnology inventions are filed as utility patents and not as plant patents. As a utility patent it is possible to protect plant genes, rather than just the plant, and to control the use of the genetic material of a number of plants and for multiple uses such as pharmaceutical, pest protection, herbicide resistance, oil production, etc.

In the USA an inventor is given a period of one year to file a patent application after disclosing the invention. The right to file for foreign patents is lost upon disclosure. This is not true elsewhere. Also, in the USA the person who invents something first is granted a patent even if a rival inventor, who invented the same thing later, files for a patent first. Elsewhere in the world the person who files first, regardless of when the invention was developed, obtains the patent.

The preparation of a patent application is quite complex and generally an attorney is required to draft and prosecute the application. Especially important is the drafting of patent claims. Claims are that portion of the patent which describes what can be accomplished with the invention and are what is protected for the life of the patent: no-one can do what the claims claim without the inventor's permission. The selection of an attorney is important, as an attorney familiar with the field of the invention can draft much better and broader claims than one not familiar with the field. Because an attorney is involved, patents cost far more than copyrights or trademarks – seldom less than US$10,000 and generally much more.

While a copyright is granted upon creation, a patent application may take more than two years to get through the US PTO. A utility patent for a mechanical device may be granted within 18 months, where it may take a biotechnology patent more than 30 months to issue.

A patent is only enforceable in the country which issues it. While a Patent Cooperation Treaty (PCT) application can provide addtional time, a separate patent application is necessary for each country in which the inventor desires protection. The cost for filing in a number of countries is great and costs can easily exceed US$100,000. If one does not get protection in a country, anyone in that country can use, manufacture and sell the invention. However, the products produced in the non-patent countries cannot be sold in the countries where the invention is protected by patent.

The patent is a document and in the USA has the following components: (i) a cover page which lists information such as patent title, patent number, date of filing, date of issue; an abstract; (ii) figures or drawings; (iii) the body of text with an introduction, specifications, examples and other background information; and (iv) the claims section which describes exactly what one can do or accomplish with the invention. A copy of a US patent is provided in Appendix 1.1 at the end of this chapter.

Copyright, trademark and patent are the basic means of protecting creations and discoveries. There are two additional means of protection, each of which has advantages over the basic methods described above. These means of protection are the trade secret and plant variety protection. These are discussed further below.

WHAT IS A TRADE SECRET?

The trade secret is probably the most interesting of the rights available in IP. A trade secret is any information that gives a company a competitive edge over competitors and which the company maintains as secret and away from public knowledge.

The US Uniform Trade Secret Act (UTSA) defines trade secret as, 'Any formula, pattern, device or compilation of information which is used in one's business, and which gives the business an advantage over competitors who do not know how to use it.'

Trade secret rights are mainly kept and enforced through agreements between employers and employees. Usually at the time employment begins, an employer makes an employee sign an agreement which grants the employer trade secret protection. The trade secrets protected under these agreements are non-patented projects involving substantial time and cost for the company and, in some cases, rejected or failed company projects. Additionally, these agreements also protect the company by preventing its competitors from enticing key personnel since these individuals cannot divulge the trade secret material without incurring severe penalties. Trade secret law provides remedies to companies who have had trade secrets stolen by competitors or employees. Criminal prosecution of an ex-employee who steals trade secrets from their employer is a recognized remedy. It is a criminal offense in the USA for a person knowingly to reveal confidential processes or formulations, which are maintained as secret information by the company, without the express consent of the employer.

The protection provided by a trade secret has an indeterminable term, which may be perpetual. The term is as long as it takes the public or a competitor to determine how to make the product and to ascertain the nature and identity of the trade secret.

The nature or the identity of a product is maintained secret for as long as the company can keep this information from becoming public knowledge. For example, the Coca Cola company has kept the formula of its base syrup flavoring a secret for many years. Another company that has kept a trade secret for a long period of time is the Polaroid company, which has kept the instant film chemical formula out of public knowledge. These two companies have closely guarded these pieces of tangible but restricted knowledge.

Trade secrets are much more common in industry where scholarly publication is not required and where the value of the information depends on how well it can be kept secret from competitors and the public. In contrast, universities and governmental laboratories are expected to share their findings through publication and presentation, making it almost impossible to maintain a trade secret.

Trade secret is sometimes the only thing that allows a company to compete in today's market. Companies spend millions of dollars on security measures to protect their trade secrets. In many instances, trade secret is the foundation of a free enterprise and marketable product.

There is no direct cost for a trade secret; however, the costs of maintaining a trade secret can be great. Costs include developing and entering into employee agreements, policing of employees and agreements, and preventing other companies from learning about the secret.

WHAT IS PLANT VARIETY PROTECTION?

Plant variety protection (PVP) allows one to protect new varieties of sexually reproducing plant varieties for a term of 20 years. There are several advantages to this type of protection over plant patents: (i) the cost is much lower, US$2500 compared with US$10,000–20,000; (ii) the simplicity of application (a breeder can complete the required form and an attorney is not needed); (iii) the requirements for protection are less than those for patenting; and (iv) the protection is quite similar.

Generally, PVP is not sought for transformed plants, i.e. plants into which genes have been incorporated through biotechnology, but for plants or varieties which have been developed through traditional breeding. These plants or varieties of crop plants are usually economically viable for five to ten years, depending on the rate of disease and pest infestation. Breeders are continually developing new varieties and a breeder may have one or more new varieties ready for release each year. The high cost of patenting would prohibit most breeders and companies from getting protection for these varieties, so the PVP avenue provides an appropriate and alternative means of protection. PVP, as with the other types of protection, is only enforceable in the country for which protection has been granted.

SUMMARY

IPs when expressed in a tangible form can be protected from unauthorized use. Literary works, including computer software, are protected by copyright; symbols and key brief phrases are protected by trademark; and inventions are protected by patent. The costs and time needed to obtain protection vary, with copyright being the least expensive (free) and quickest (immediate) and patent being the most expensive (thousands of dollars) and slowest (may take more than three years). Two other types of protection are available. One is trade secret, which means what it says – keeping the IP a secret. The other is PVP, which provides an adequate and inexpensive means to protect plants. Protection for all, except trade secret, is only applicable in the country for which one has applied for protection. Violation of any of the means of protection is subject to various types of punishment including fines and imprisonment.

APPENDIX 1.1.

The US patent shown here is owned by Michigan State University. The various parts of the patent are marked with numbers to illustrate the information provided in the patent document. The key to these numbers is given below.

Cover page

1, patent number; 2, patent number in bar code; 3, date of issue of patent; 4, last name of inventor; 5, title of patent; 6, inventor's full name and city of residence; 7, the owner of the patent; 8, date on which the patent application was filed; 9, number assigned to the patent application; 10, patents against which the patent application was compared by the patent examiner to check that the invention was novel; 11, the name of the individual who examined the patent in the US PTO; 12, the name

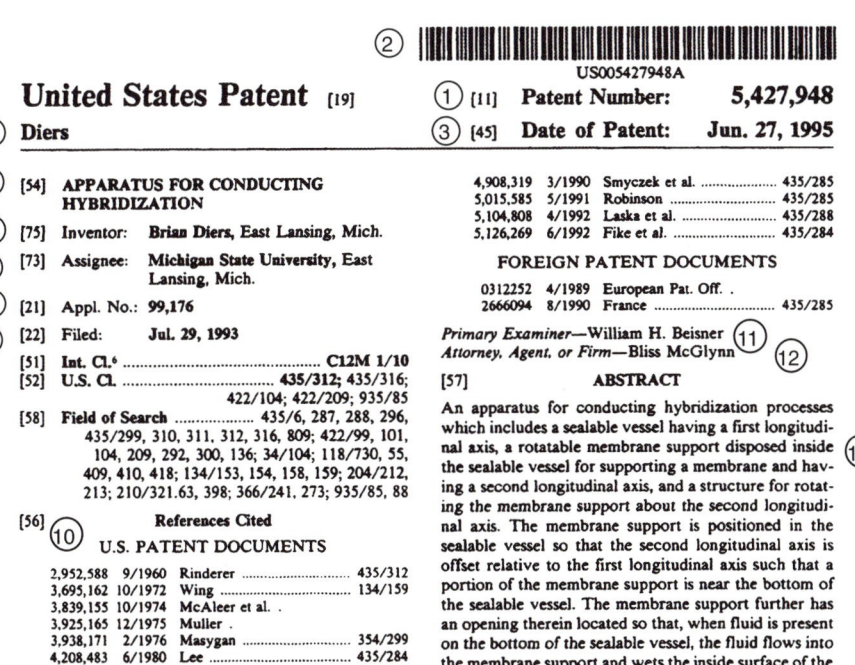

of the university's patent attorney; 13, the summary or abstract of the patent; 14, listing of the number of claims and drawing sheets.

Drawing sheet

This gives a diagram of the invention showing the various components. A brief description of the drawings is in the text of the patent. The numbers are found throughout the text of the patent.

U.S. Patent   **June 27, 1995**   **5,427,948**

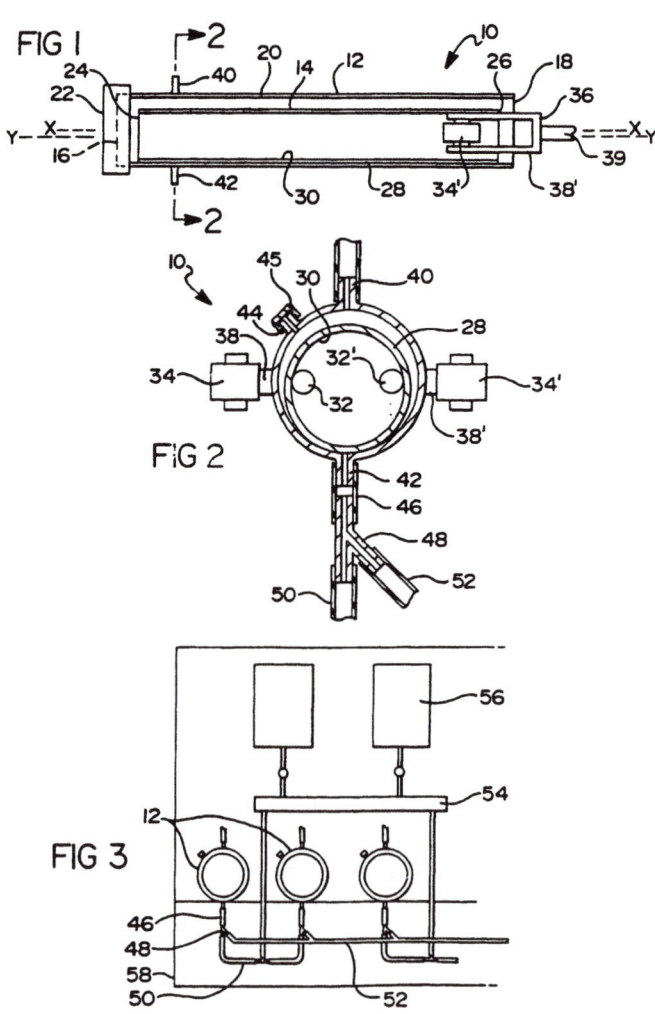

Text

This has four sections: (1) background information; (2) a more complete summary of the invention (as compared with the abstract); (3) a description of the invention with references to the drawings (note that the patent text is not presented in the usual page setup, as each column of text is numbered rather than each page); (4) The claims section (each claim is numbered).

MICHIGAN STATE PATENT 5,427,948

1

APPARATUS FOR CONDUCTING HYBRIDIZATION

BACKGROUND OF THE INVENTION

1. Field of the Invention

The present invention relates generally to apparatuses for conducting micro-biological experiments, and, more particularly, to apparatuses for conducting hybridization processes.

2. Description of the Related Art

DNA fragments are used in hybridization reactions to characterize the identity of specific genes and to study the molecular organization of genomic sequences. Specific hybridization techniques have been developed in the past, namely, the Southern hybridization process, the Northern hybridization process, and the Western hybridization process.

In the Southern hybridization process, single-stranded DNA fragments are bound to a membrane and the bound DNA fragments are then hybridized by exposure to a hybridization solution containing radioactive materials. The single-stranded DNA fragments bound to the membrane that are complementary to the radioactive material will form hybrids. The membranes are then washed and ready for analysis.

Past procedures for conducting the Southern hybridization process have involved tedious washing steps and potential human exposure to radioactive materials.

SUMMARY OF THE INVENTION

It is, therefore, one object of the present invention to provide an apparatus for conducting hybridization processes which simplifies the washing step.

It is another object of the present invention to provide an apparatus for conducting hybridization processes which uses a minimum amount of hybridization solution.

It is yet another object of the present invention to provide an apparatus for conducting hybridization processes which reduces human exposure to radioactive materials.

To achieve the foregoing objects, the present invention is an apparatus for conducting hybridization which includes a sealable vessel having a first longitudinal axis, a rotatable membrane support disposed inside the sealable vessel for supporting a membrane and having a second longitudinal axis, and means for rotating the membrane support about the second longitudinal axis. The membrane support is positioned in the sealable vessel so that the second longitudinal axis is offset relative to the first longitudinal axis such that a portion of the membrane support is near the bottom of the sealable vessel. The membrane support further has an opening therein located so that, when fluid is present on the bottom of the sealable vessel, the fluid flows into the membrane support and wets the inside surface of the membrane support while it is being rotated.

One advantage of the present invention is that an apparatus is provided for conducting hybridization processes. Another advantage of the present invention is that the hybridization apparatus provided simplifies the washing step of the hybridization process so that the washing step is faster, easier, and less labor intensive. Another advantage of the present invention is that the hybridization apparatus uses a minimum amount of hybridization solution. Yet another advantage of the present invention is that the hybridization apparatus minimizes human exposure to radioactive materials.

2

present invention is that the hybridization apparatus minimizes human exposure to radioactive materials.

Other objects, features, and advantages of the present invention will be readily appreciated as the same becomes better understood after reading the subsequent description taken in conjunction with the appendant drawings.

BRIEF DESCRIPTION OF THE DRAWINGS

FIG. 1 is a side elevational view of an apparatus for conducting hybridization processes according to the present invention.

FIG. 2 is a sectional view taken along line 2—2 of FIG. 1.

FIG. 3 is a front elevational view of a series of apparatuses of FIG. 1 illustrated in operational relationship with an incubator.

DESCRIPTION OF THE PREFERRED EMBODIMENT(S)

Referring to FIG. 1, an apparatus 10 for conducting hybridization processes is shown. Apparatus 10 includes sealable vessel 12 and membrane support 14 disposed inside and resting on the bottom of sealable vessel 12. Sealable vessel 12 and membrane support 14 may be formed of glass or plastic or other material suitable for hybridization techniques. Although sealable vessel 12 is shown as being cylindrically-shaped, other suitable shapes may be used in the invention. Sealable vessel 12 has open front end 16, closed back end 18, cylindrical wall 20 connecting front end 16 and back end 18, and longitudinal axis X—X which extends from the center of front end 16 to the center of back end 18. Sealable vessel 12 also has cap 22 closing front end 16. Cap 22 may be, e.g., a screw cap or a press fit cap. Preferably, cap 22 includes sealing means (not shown), such as a gasket, to provide optimum sealing. Open front end 16 and cap 22 allows the opening of sealable vessel 12 so that membrane support 14 may be placed therein.

Membrane support 14 has open front end 24, open back end 26, and continuous cylindrical wall 28 connecting front end 24 and back end 26. Cylindrical wall 28 has inside surface 30, best seen in FIG. 2, for supporting a membrane (not shown). Typically, during the hybridization process, a rectangular membrane (not shown), formed of, e.g., nitrocellulose or a nylon derivative, is rolled into a tubular shape and placed inside membrane support 14 so that it lays against inside surface 30. It is preferred that cylindrical wall 28 be substantially continuous to provide maximum support for such a membrane.

Membrane support 14 is positioned in sealable vessel 12 so that a portion of cylindrical wall 28 is near the bottom of sealable vessel 12. To accomplish this, membrane support 14 rests on the bottom of sealable vessel 12. Alternatively, membrane support 14 may be secured by any suitable means to maintain a position near the bottom of sealable vessel 12.

Membrane support 14 is rotatable about longitudinal axis Y—Y which extends from the center of front end 24 to the center of back end 26. In addition, membrane support 14 has an outer diameter which is smaller than the inner diameter of sealable vessel 12, which minimizes frictional contact between membrane support 14 and sealable vessel 12 during rotation of membrane support 14. Since membrane support 14 has an outer diameter which is smaller than the inner diameter of sealable vessel 12 and membrane support 14 rests on the

Continued

MICHIGAN STATE PATENT 5,427,948

3

bottom of the inside surface of sealable vessel 12, longitudinal axis Y—Y of membrane support 14 is offset from and below longitudinal axis X—X of sealable vessel 12.

As illustrated in FIGS. 1 and 2, apparatus 10 includes first magnets 32 and 32' attached to inside surface 30 of membrane support 14 which are moved by second magnets 34 and 34', respectively, outside sealable vessel 12. Magnets 32 and 32' are attached near back end 26 of membrane support 14 and are positioned 180° apart along the same radius of membrane support 14. It is preferred that the inside surface of membrane support 14 be large enough so that a membrane as described above may be placed on the smooth inner surface without contacting a magnet. Laying the entire membrane against inner surface 30 avoids the formation of air bubbles on the membrane surface which would lead to non-uniform hybridization. Magnets 32 and 32' may be attached to inside surface 30 by any suitable means, such as by adhesive or screws.

Magnets 34 and 34' are positioned outside sealable vessel 12 and near magnets 32 and 32', respectively. Therefore, magnets 34 and 34' are spaced 180° apart and are axially and radially in line with magnets 32 and 32'. Magnets 34 and 34' are held by retainer 36 which has arms 38 and 38', to which magnets 34 and 34' are attached, respectively. Retainer 36 is connected to drive shaft 39 which is rotatable by a motor (not shown). The rotation of drive shaft 40 causes retainer 36 to rotate which, in turn, causes magnets 34 and 34' to rotate about the circumference of sealable vessel 12. The rotation of magnets 34 and 34' causes the rotation of magnets 32 and 32', thereby causing the rotation of membrane support 14. It should be appreciated that, although other means for rotating membrane support 14 may be used, the magnets described herein are desirable as they present minimal parts which could lead to leakage problems.

In an alternative embodiment, a single magnet may be attached to membrane support 14 which is rotated by a single magnet which rotates about the circumference of sealable vessel 12. As another variation of the invention, sets of more than two magnets may be employed. As another way of rotating membrane support 14, membrane support 14 may be attached to a drive shaft which is rotated by a motor.

Sealable vessel 12 further has vent port 40 near the top of the vessel, conduit port 42 near the bottom of the vessel, and insertion port 44 on the upper portion of the vessel. Vent port 40 provides fluid communication between the outside of sealable vessel 12 and the inside of sealable vessel 12 and is present to allow air to vent from the vessel when adding fluid to sealable vessel 12 and to allow air to enter the vessel when removing fluid from sealable vessel 12. Insertion port 44 has screw top 45 thereon and is present to provide access into sealable vessel 12 for addition of hybridization solution and the like without having to remove cap 22 from sealable vessel 12.

Conduit port 42 is present to provide fluid communication between the outside of sealable vessel 12 and the inside of sealable vessel 12. Conduit 46 is shown attached to sealable vessel 12 in FIGS. 2 and 3. A valve may be installed in conduit 46 to control fluid flow into and out of sealable vessel 12. Also shown in FIG. 3 are several sealable vessels 12 placed in parallel so that several hybridization reactions can take place at one time. Each sealable vessel 12 can have a different type of membrane, allowing for the testing of several mem-

4

branes at a time. "Y" connector 48, connected to conduit 46, provides means for attaching conduits 50 and 52 to conduit 46. Conduit 50 is connected to manifold 54 which is, in turn, fluidly connected to fluid source 56 which may contain, e.g., hybridization solution or wash solution. Conduit 52 may lead to a waste container (not shown) into which fluid from sealable vessel 12 may drain.

The sealable vessels 12 shown in FIG. 3 are placed in incubator 58 which maintains the temperatures of the vessels and their contents at the desired temperature.

During operation of apparatus 10, first a DNA-bound membrane is rolled into tubular form and placed inside membrane support 14. Membrane support 14 is then placed inside sealable vessel 12, and cap 22 is secured onto sealable vessel 12. Hybridization solution is then added into sealable vessel 12 through either insertion port 44 or through conduit 46. Typically, hybridization solution is added to fill less than about ½ the volume of sealable vessel 12. Hybridization solution enters membrane support 14 through open front and back ends 24 and 26 and wets inside surface 30 of membrane support 14, thereby wetting the membrane therein.

Membrane support 14 is then caused to rotate by rotating magnets 34 and 34' around the circumference of sealable vessel 12. The rotation of membrane support 14 causes the continuous wetting of the entire inside surface 30 and of the entire membrane. The rotation of membrane support 14 and its proximity to the bottom of sealable vessel 12 allows the reaction to take place with a minimum amount of hybridization fluid.

Open ends 24 and 26 of membrane support 14 allow the fluid to enter the membrane support. Alternatively, only one open end or openings in other locations on membrane support 14 would serve the same purpose.

The hybridization reaction is allowed to occur for the desired length of time in the incubator. Once the hybridization is completed to the extent desired, the spent hybridization solution is drained from sealable vessel 12 through conduit 52 and into a waste container. The membranes are then washed by filling the sealable vessels with wash solution to any desired level. Preferably, the wash solution is from fluid source 56. From fluid source 56, the wash solution flows through manifold 54, through conduit 50, through "Y" connector 48, through conduit 46, and into sealable vessel 12. Membrane support 14 is rotated to assist in the washing process. Used wash solution may be drained from sealable vessel 12 and into a waste container. The washing procedure may be repeated as desired. After washing, the membranes may be removed from the apparatus and analyzed, as desired. It can be seen that, using apparatus 10 of the present invention, the washing of the membranes is accomplished without handling the membranes or any radioactive materials.

Accordingly, apparatus 10 of the present invention is useful for hybridization processes, such as Southern, Northern and Western hybridization techniques. Apparatus 10 used for conducting hybridization processes simplifies the washing step, uses a minimum amount of hybridization solution, and reduces human exposure to radioactive materials.

The present invention has been described in an illustrative manner. It is to be understood that the terminology which has been used is intended to be in the nature of words of description rather than of limitation.

Many modifications and variations of the present invention are possible in light of the above teachings.

Continued

MICHIGAN STATE PATENT 5,427,948

5

Therefore, within the scope of the appended claims, the present invention may be practiced otherwise than as specifically described.

What is claimed is:

1. An apparatus for conducting hybridization processes, comprising:

    (a) a cylindrically-shaped sealable vessel having a first longitudinal axis;

    (b) a cylindrically-shaped rotatable membrane support disposed inside said sealable vessel for supporting a membrane and having a second longitudinal axis, said membrane support having a front end and a back end and a cylindrical wall connecting said front end and said back end, said cylindrical wall having an inside surface for supporting the membrane longitudinally therealong in continuous contact with said inside surface, said membrane support being positioned in said sealable vessel so that the second longitudinal axis is offset relative to the first longitudinal axis such that a portion of said membrane support is near the bottom of said sealable vessel, said membrane support having an opening therein located so that, when fluid is present on the bottom of said sealable vessel, the fluid flows into said membrane support and wets the inside surface of said membrane support while it is being rotated; and

    (c) means for rotating said membrane support about the second longitudinal axis.

2. An apparatus as set forth in claim 1, further comprising a conduit port attached to the sealable vessel providing fluid communication between the outside of the sealable vessel and the inside of the sealable vessel.

3. An apparatus as set forth in claim 1, further comprising a vent port near the top of the sealable vessel providing fluid communication between the outside of the sealable vessel and the inside of the sealable vessel.

4. An apparatus as set forth in claim 1, wherein the sealable vessel has a front end and a back end and a wall connecting the front and back ends.

5. An apparatus as set forth in claim 1, further comprising means for opening the sealable vessel so that the membrane support may be placed therein.

6. An apparatus as set forth in claim 1, wherein the membrane support rests on the bottom of the sealable vessel.

7. An apparatus as set forth in claim 1, wherein the means for rotating the membrane support comprises a first magnet attached to the membrane support, a second magnet axially in line with the first magnet and outside the sealable vessel, and means for rotating the second magnet around the circumference of the sealable vessel to thereby rotate the first magnet about the second longitudinal axis.

8. An apparatus as set forth in claim 1, wherein said cylindrical wall of said membrane support is continuous.

9. An apparatus as set forth in claim 1, wherein the ends of said membrane support are open ends.

10. An apparatus for conducting hybridization processes, comprising:

    (a) a cylindrically-shaped sealable vessel having a front end and a back end and walls connecting said front and back ends and having a first longitudinal axis;

    (b) a cylindrically-shaped rotatable membrane support disposed inside said sealable vessel, said membrane support having an open front end and an

6

open back end and a cylindrical wall connecting said open front and back ends, said cylindrical wall having an inside surface for supporting a membrane longitudinally therealong in continuous contact with said inside surface, said membrane support having a second longitudinal axis, the second longitudinal axis offset relative to the first longitudinal axis such that said cylindrical wall rests on the bottom of said sealable vessel, said membrane support having said open front and back ends so that when fluid is present on the bottom of said sealable vessel, the fluid flows into said membrane support, and wets said inside surface of said membrane support while it is being rotated;

    (c) means for rotating said membrane support about the second longitudinal axis;

    (d) a conduit port near the bottom of said sealable vessel providing fluid communication between the outside of said sealable vessel and the inside of said sealable vessel;

    (e) a vent port near the top of said sealable vessel providing fluid communication between the inside of said sealable vessel and the outside of said sealable vessel; and

    (f) means for opening said sealable vessel so that said membrane support may be placed therein.

11. An apparatus as set forth in claim 10, wherein cylindrical wall of the membrane support is continuous.

12. An apparatus as set forth in claim 10, wherein means for rotating the membrane support comprises a first magnet attached to the cylindrical wall of the membrane support, a second magnet axially in line with the first magnet and outside the sealable vessel, and means for rotating the second magnet around the circumference of the sealable vessel to thereby rotate the first magnet about the second longitudinal axis.

13. An apparatus for conducting hybridization processes, comprising:

    (a) a cylindrically-shaped sealable vessel having a front end and a back end and walls connecting said front and back ends and having a first longitudinal axis;

    (b) a cylindrically-shaped rotatable membrane support inside said sealable vessel resting on the bottom thereof, said membrane support having an open front end and an open back end and a continuous cylindrical wall connecting said open front and back ends, said cylindrical wall having an inside surface for supporting a membrane longitudinally therealong in continuous contact with said inside surface, said membrane support having a second longitudinal axis offset relative to the first longitudinal axis and having said open front and back ends so that when fluid is present on the bottom of said sealable vessel, the fluid flows into said membrane support, and wets said inside surface of said membrane support while it is being rotated; and

    (c) a first magnet attached to said cylindrical wall of said membrane support, a second magnet axially in line with said first magnet and outside said sealable vessel, and means for rotating said second magnet around the circumference of said sealable vessel to thereby rotate said first magnet about the second longitudinal axis;

    (d) a conduit port near the bottom of said sealable vessel providing fluid communication between the outside of said sealable vessel and said inside of the sealable vessel;

Continued

MICHIGAN STATE PATENT 5,427,948

7

(e) a vent port near the top of said sealable vessel providing fluid communication between the inside of said sealable vessel and the outside of said sealable vessel; and

(f) means for opening said sealable vessel so that said membrane support may be placed therein.

14. An apparatus for conducting hybridization processes, comprising:

(a) a cylindrically-shaped sealable vessel;

(b) a cylindrically-shaped rotatable membrane support disposed inside said sealable vessel and having a continuous cylindrical wall having an inside surface for supporting a membrane longitudinally therealong in continuous contact with said inside surface and having a longitudinal axis, said membrane support being positioned in said sealable vessel so that said membrane support rests on the bottom of said sealable vessel, said membrane support having an opening therein located so that, when fluid is present on the bottom of said sealable vessel, the fluid flows into said membrane support and wets the inside surface of said membrane support while it is being rotated; and

(c) means for rotating said membrane support about the longitudinal axis.

15. An apparatus as set forth in claim 14, further comprising a conduit port attached to the sealable ves-

8

sel providing fluid communication between the outside of the sealable vessel and the inside of the sealable vessel.

16. An apparatus as set forth in claim 14, further comprising a vent port near the top of the sealable vessel providing fluid communication between the outside of the sealable vessel and the inside of the sealable vessel.

17. An apparatus as set forth in claim 14, wherein the sealable vessel has a front end and a back end and a wall connecting the front and back ends.

18. An apparatus as set forth in claim 14, further comprising means for opening the sealable vessel so that the membrane support may be placed therein.

19. An apparatus as set forth in claim 14, wherein the membrane support has an open end at each end of the cylindrical wall.

20. An apparatus as set forth in claim 14, wherein the means for rotating the membrane support comprises a first magnet attached to the cylindrical wall of the membrane support, a second magnet axially in line with the first magnet and outside the sealable vessel, and means for rotating the second magnet around the circumference of the sealable vessel to thereby rotate the first magnet about the longitudinal axis of the membrane support.

* * * * *

Acquiring Protection for Improved Germplasm and Inbred Lines

2

John H. Barton

Stanford University, Crown Quadrangle, Stanford, CA 94305, USA

INTRODUCTION

This chapter explores the intellectual property issues involved in traditional breeding and in moving from natural material to the improved lines that are marketed themselves or used as parents of a hybrid. The chapter begins with a review of access to unimproved germplasm and the implications of the United Nations Convention on Biological Diversity. It then considers relevant forms of intellectual property protection as applied in the USA. These include the plant variety protection (PVP) system, the regular patent system and trade secrecy. The chapter concludes with a description of enforcement.

THE INTERRELATIONSHIPS BETWEEN INTELLECTUAL PROPERTY AND BIODIVERSITY

Ultimately, much of the agricultural germplasm of the world comes from the developing nations. It was, for example, Mexico in which corn (maize) was domesticated and the Andes in which the tomato and the potato were domesticated. It is the developing nations, too, that contain wild relatives or land races, sometimes incorporating resistances and other characteristics that may be of interest to a contemporary plant breeder.

At one time, the scientific norm was to collect germplasm freely in any nation, including developing nations, and to use it in breeding. Moreover, as the world's gene banks were organized during the 1970s, the collections were made on a similar basis – the return to the source nations would be through the benefits of the improved varieties developed with the assistance of the collected material. But during the 1970s, the developed nations moved quite strongly to adopt PVP, a form of intellectual property protection on plants to be discussed below. There arose concerns, based on the perception that it was unfair for the source material contributed by the developing nations to be transferred freely, while breeding activities contributed by developed nations were being rewarded with intellectual property rights.

These concerns led to political movements within the Food and Agriculture Organization (FAO), which created a Commission on Plant Genetic Resources and passed an International Undertaking on Plant Genetic Resources in 1983 (FAO, 1983). They also became major factors shaping the United Nations Convention on Biological Diversity, signed at Rio de Janeiro in 1992. Continued negotiations are looking towards, for example, recognition of a right of the small farmer who has contributed to the genetic resource through the selection of seeds over the generations, creation of a fund for compensating source nations for past transfers of materials, and development of a mechanism to provide some form of source nation right in materials contained in gene banks at the time of the Convention on Biological Diversity.

The Convention on Biological Diversity itself includes carefully negotiated provisions governing genetic resources, as part of a much broader package oriented towards conservation of biological diversity in its natural habitats and in collections. The Convention's Article 19 affirms the sovereign rights of nations over their genetic materials, but leaves it clear that those genetic materials that were earlier transferred out of their nation of origin have entered the public domain and can be used freely for any purpose (Barton, 1992). (Thus, certain of the current proposals for new international arrangements are efforts to modify the understanding reached at Rio de Janeiro.)

The clear implication is that, in general, no further genetic material will be collected from any developing nation except pursuant to a 'material transfer agreement' (MTA). The MTA will be agreed between the collector and appropriate national authorities, and will govern the arrangements under which the material is transferred. These may include an allocation of profits or a provision that the material cannot be used commercially without a further agreement allocating profits. There may also be provisions that, for example, restrict the acquisition of intellectual property rights on the material, and there will normally be a prohibition of transfer of the material without building a chain of responsibility. Not all nations have yet adopted the legislation needed to enforce this right that they hold under the Convention. Moreover, some nations, looking to the costs of preparing and implementing these agreements and looking to the benefits of free exchange of genetic material, may choose not to require restrictive MTAs. But the current trend is toward restriction of free flow, and it will sometimes be necessary for a breeder to work with source nations in order to ensure good title to the material used in a breeding program.

PLANT VARIETY PROTECTION

There are two significantly different regimes for the protection of plant breeding materials: the PVP (plant variety protection or plant breeders' rights system) and the regular patent system. For general reviews of the application of these systems to plant agriculture, see Baenziger *et al.* (1993), Hamilton (1993a), Parr (1993) and Roberts (1996).

The regime designed specifically for traditional plant breeding is the PVP system. It is designed to give these breeders an increased incentive to develop new varieties while respecting their traditions of exchanging material. The US version, passed in 1970 and since updated (7 USC §§2321–2582), grants protection to varieties that are 'new,' 'distinct,' 'uniform' and 'stable' (7 USC §2402). To be new, the variety must not have been sold previously, although there is a grace period of one year, and longer for foreign use. Distinctness requires that the variety be clearly distinguishable from previous varieties – this is not as severe an inventive step requirement as is typical of patent law. Uniformity requires that any variations be 'describable, predictable, and commercially acceptable'. Stability requires that, when reproduced, the variety 'remain unchanged with regard to [its] essential and distinctive characteristics … with a reasonably degree of reliability.' Moreover, seeds of the variety must be deposited (7 USC §2422).

The PVP law applies to sexually reproduced plants and tubers. There was an earlier law, the Plant Patent Act of 1930 (35 USC §§161–164), that applies to varieties that propagate asexually, and is applied by the Patent Office, which can consult with the Department of Agriculture (27 CFR §1.167).

Protection under PVP is by means of a certificate granted by an office of the Department of Agriculture upon receipt of a relatively simple and inexpensive application. The variety must be given a name (7 USC §2422), and this name, of course, becomes an important part of the marketing of the variety, and may be given trademark protection as well. Protection is for 20 years, or 25 years in the case of a tree or vine (7 USC §2483). The certificate entitles its holder to be the exclusive marketer of the relevant variety, and also of the product of the variety. This right may, of course, be licensed to others. The certificate does not, however, prevent others from using the variety in efforts to breed further varieties, nor does it prevent farmers from reusing harvested material (7 USC §2541). Farmers had at one time also been able to sell their seed under some circumstances (Asgrow Seed Co. v. Winterboer, 513 US 179 (1994)); this right was significantly narrowed in the 1994 revision of the act (PL no. 103–349, 6 October 1994).

The PVP laws of various nations are harmonized through an international treaty, e.g. UPOV (1978, 1991) (named after the French language acronym for the International Union for the Protection of New Varieties of Plant). This treaty establishes standards for PVP legislation and requires its parties to offer one another's breeders the opportunity to obtain PVP certificates just as if they were nationals. Under the older versions of this treaty (e.g. UPOV, 1978), nations were required both to allow use of protected materials for breeding of additional new varieties, and to allow farmers to reuse their harvest for seed purposes. Article 15 of the new (1991) version, which is likely to come into force in 1998, permits nations to allow farmers to reuse seed, but does not require them to do so. As noted above, the USA has made this authorization. Article 14 of this new version adopts a concept of 'essentially derived variety', a concept implemented at 7 USC §2541. A breeder remains free to use a protected variety and to make any change in such a variety, but is subject to the rights of the owner of the initial variety if that change is so small as to leave the new variety 'essentially derived'. Examples listed in this Article are varieties made 'by the selection of a natural or induced mutant, or of a somaclonal variant, the selection of a variant individual from plants of the initial variety, backcrossing, or transformation by genetic engineering'.

There is strong evidence that adoption of a PVP system in the USA increased private sector plant breeding (Butler and Marion, 1985), and the rise of biotechnology-based breeding offers no reason to question this judgment. It is also clear, however, that PVP does not provide adequate protection for a firm which has sequenced an important gene and transformed plants with it. If PVP were the breeder's only protection, another breeder could purchase the protected material and breed the gene into a new variety. This is in no way an infringement of PVP rights, but it clearly significantly decreases the market position of the initial breeder.

THE REGULAR PATENT SYSTEM

For this and many other reasons, biotechnology-oriented breeders have turned to the regular patent system. After initial hesitation, surmounted by Diamond v. Chakrabarty (447 US 303 (1980)), the US Patent and Trademark Office began to issue many different types of regular patents protecting biotechnological methods of breeding and biotechnologically produced plants.

Patent system concepts

As will be recalled from Chapter 1, an invention or discovery must be novel, non-obvious, useful and enabled, in order to be patentable. 'Novelty' means that the invention has not been anticipated by publication or use in the market (35 USC §102). (Unlike most nations, the USA allows a one-year grace period between the time of a publication and the time at which a patent can be filed.) 'Non-obviousness' means that the invention is an actual advance in the state of the art. The US definition is that a patent shall be denied if 'the subject matter as a whole would have been obvious at the time the invention was made to a person having ordinary skill in the art to which said subject matter pertains' (35 USC §103). Likewise, the standard of 'utility' (35 USC §101) is intended as one way to distinguish basic scientific advances from patentable inventions. 'Enablement' means that the patent describes a way to carry out the invention, typically through a description in the patent (35 USC §112). Sometimes enablement may also require deposit of actual genetic material, e.g. a seed, when this line cannot be reliably produced on the basis of a written description. This seed must be available to the public once the patent enters into force (37 CFR §1.808). Such deposit can be made at any of a number of institutions and there is an international treaty allowing each nation to recognize deposits in other nations (Budapest Treaty, 1977). Under some circumstances, enablement may require presentation of gene or amino acid sequences; this sequence must be provided in machine-readable form (37 CFR §§1.821ff).

The patent itself includes both a description of how to practise the invention and a statement of claims, which precisely define the exclusive rights conferred by the patent. In evaluating the possibility of infringement, it is these claims that must be consulted. Obtaining a patent is both slower and more expensive (typically US$20,000 for legal costs and filing fees) than obtaining a PVP certificate; expenses of global coverage can easily rise into the hundreds of thousands of US dollars. The term of protection is 20 years from the date of application, with the possibility of extension in the event of certain delays (35 USC §154).

Varieties of patents

The Board of Patent Appeals and Interferences of the US Patent and Trademark Office has interpreted Diamond v. Chakrabarty to mean that any plant can be patented, provided it satisfies the basic standards for intellectual property. In particular, it has concluded that the availability of a special PVP system for plants does not exclude patentability under the regular patent laws (*Ex parte* Hibberd, 227 USPQ 443 (1985);

Anon., 1985). It would be very difficult to read Chakrabarty in any other way. Although there had been some debate about the desirability of such 'double protection', it has thus become generally assumed in the USA that one can obtain both a patent and a PVP certificate for the same organism.

In the USA, it is possible to obtain a patent on a gene and its application in a plant and on basic processes and inventions in the way discussed in the previous chapter of this book. We will note these possibilities very briefly, and then turn to the protections available on a plant or inbred line itself. The patent on a gene and on transformed plants utilizing the gene is frequently written with a number of claims covering, for example: an isolated or purified protein, the isolated or purified nucleic acid sequence that codes for the protein, plasmids and transformation vectors containing the gene sequence, plants (or seeds for such plants) transformed with such vectors and containing the gene sequence, and the progeny (or seeds) of such plants. For an example that shows a number of these claims, see Zaitlin *et al.* (1997). This structure of the claims, which reach isolated versions of the gene or protein, protects the patent holder against use of the gene by another biotechnologist, but leaves anyone free to use and breed with organisms containing the gene naturally. Another category of patents covers basic processes and inventions. Here, there are many extremely important patents, e.g. on transformation processes, promoters, the use of virus coat proteins to confer resistance, and antisense technology.

It is also possible in the USA to obtain claims covering broad groups of transgenic plants, as exemplified by the *Agracetus* patents on all transgenic cotton (Umbeck, 1992). The breadth of such a patent is extremely significant and has been the subject of severe criticism (Stone, 1995). The underlying legal issue is enablement; the claims are supposed to reach as far as the disclosure enables a person of ordinary skill in the art to do the claimed action without 'undue' experimentation. When a person applies for a patent after transforming several strains of a species with several different genes, there is an obvious question as to whether that person has actually enabled transformation of *all* strains with *all* genes. Although it is likely that no-one knows the answer to this question at the time of patent application, the burden of proof in the USA on this issue is on the patent office to show that a claim was not enabled. Comparable issues are posed by claims based on plant descriptions, for example, 'a hybrid maize plant characterized by a genetic factor which confers an extra leaf phenotype, said genetic factor being capable of transmission to progeny substantially as a single dominant gene' (Muirhead and Shaver, 1985). As with the cotton patent, there is the question whether the disclosure of one or several lines with the particular characteristics should give rights over all such lines.

Finally, there are the patents of most importance to this chapter, those on a specific variety. Although the validity of this form of patent has not yet been tested in court, it has become normal practice to consider regular patent protection for a variety as a reasonable alternative or supplement to PVP and trade secrecy (for inbred lines used as parental hybrids). If this technique is successful, it can be used to protect against a farmer's seed reuse and against breeders seeking to use the material. This use of the regular patent system may thus provide a way to avoid the limitations of the PVP system.

The claims in a variety patent will specify a variety by its name or by a designation, for example:

> **1.** Seed of maize inbred line designated PHDG1 and having ATCC [American Type Culture Collection] Accession No 97663.
> **2.** A maize plant and its parts produced by the seed of claim 1 and its plant parts.
>
> (Piper, 1997)

The claims may cover inbred lines or hybrids; they may cover seeds or plants; and they may attempt to extend to progeny. The patent just cited goes on:

> **10.** A method for producing first generation (F_1) hybrid maize seed comprising crossing a first inbred parent maize plant with a second inbred parent maize plant and harvesting the resultant first generation (F_1) hybrid maize seed, wherein said first or second parent maize plant is the maize plant of claim 2.
> **13.** An F_1 hybrid seed and plant produced by the method of claim 10.

Another approach to claiming progeny is 'A hybrid corn plant, wherein at least one ancestor of said hybrid corn plant is the corn plant [of the claimed inbred line]' (Strissel *et al.*, 1992). For general discussion of such claims and other examples, see Seay (1993).

The evaluation of obviousness in such patents is quite difficult. In *Ex parte C*, 27 USPQ.2d 1492 (Board of Patent Appeals and Interferences, 1992), the Board appeared to assume that breeding of a new soybean variety could provide the basis for a regular patent, but did not accept the mere fact of difference from previous varieties as adequate:

> We have reviewed the data and the declaration but are unpersuaded of patentability because there is nothing of record which explains why the differences between the claimed variety and a rot resistant variety such as 'Pella 86' are so significant and unexpected that they should weigh more heavily than the numerous similarities between the claimed variety and the varieties of the cited prior art.
>
> 27 USPQ.2d at 1497

More recently, however, the Federal Circuit (a higher review group) held in *In re Sigco Research*, 36 USPQ.2d 1380 (Federal Circuit, 1995), that it was *not* obvious to apply conventional plant breeding techniques to obtain true-breeding sunflower plants whose oil had an oleic acid level of 'approximately 80% or greater'.

In order for this approach to work for the breeder interested in preventing farmers from reusing the seed, it is essential that, with an appropriate claim, it will be possible to control use of the progeny of the plant. This judgment requires an interpretation of two doctrines. One is the doctrine of patent exhaustion – in general, once a patented product is sold, the purchaser is free to use it in any way and has, in effect, an implied license for using the product, reselling it, etc. The other doctrine is that replication of an invention is an infringement. Although the issue has not yet been decided in court, the expectation among US intellectual property experts is that the exhaustion doctrine will be interpreted in such a way as to uphold a patentee's rights against a purchaser's use of the seed deriving from a patented variety. It has already been recently narrowed to uphold a patent holder's restriction of use of a medical device to a single use (Mallinckrodt Inc. v. Medipart Inc., 24 USPQ.2d 1173 (Federal Circuit, 1992).

For a breeder, another important issue is whether such claims can be effective in preventing a third party from using the inbred line as a parent or crossing a variety with an inserted gene into a different variety and marketing that variety. In other words, will they be effective in overriding the PVP principle that another breeder is free to use protected material? The answer to this question is significantly less clear. Clearly, there is no control against using the material for breeding purposes unless the claims cover that use. Thus, a claim for a specific seed or a plant would seem not to prohibit crossing of the seed or the plant with another line – the new seed and plant are not within the claims of the patent. On the other hand, if the claims of the patent include use of the material as the parent of anything else, there is at least a prima facie argument that breeding is prohibited. One counter-argument is that, as will be noted in connection with restrictive license clauses, there is a strong policy that a purchaser of material in commerce has the right to study and 'reverse engineer' it in order to ensure that scientists and technologists are able to build on and improve one another's work. A counter-argument less likely to be effective is that the use is within the 'experimental use' exemption to patent infringement. This is a court-made exemption designed in the first instance to permit academic use of an invention. Although its exact scope is unclear, except in one specific context where there has been legislation (35 USC §271(e)(1) permitting experimental work with patented pharmaceuticals in preparation for entering the market at the time the patent expires), it is generally interpreted as applying only to academic research and not to commercial research (Bruzzone, 1993; Eisenberg, 1989).

TRADE SECRECY

One of the most important forms of intellectual property protection is the trade secret system, a combination of legal principles of contract law and of legal principles against misappropriation of another's information. The contractual component recognizes and encourages private enforcement of contracts designed to protect information, e.g. confidentiality agreements between a firm and its employees. The misappropriation components protect the holder of a trade secret against, for example, one who comes into the laboratory and secretly copies laboratory notebooks. To benefit from trade secret protection, a bit of information (which can include genetic material) must 'derive independent economic value' from 'not being generally known', and 'be the subject of efforts that are reasonable under the circumstances to maintain its secrecy' (Uniform Trade Secret Act §1(4)). The effective term of the protection is as long as the secret is valuable and secret, rather than being limited to a fixed term as with the patent and PVP systems.

This body of law provides a technique for control of inbred lines used as parents of a hybrid. These lines need not be released publicly in order for the hybrid to be marketed. They can be protected through a combination of physical protection of the materials themselves and of contracts with employees and those involved in producing seed. This does not, however, prevent a third party from attempting to reconstruct the parental lines from the marketed hybrid.

Firms are therefore attempting to supplement PVP and patent protection by using contractual provisions to prohibit 'reverse engineering' of the material they sell to farmers. When one buys the seeds, the label or the reverse of the sale bill contains a restrictive provision, whose key relevant language is, for example:

> Purchaser hereby acknowledges and agrees that the production from the ... [s]eeds herein sold will be used only for feed or processing and will not be used or sold for seed, breeding, or any variety improvement purpose,
>
> Stine language, quoted in Hamilton (1993b)

The legal effectiveness of this approach is subject to debate. First, there is a question of whether this mechanism of achieving contract agreement is effective, and there are cases on both sides of the issue in such contexts as warranty disclaimers on herbicides. Moreover, as noted above, there has been a tradition in US law that one has a right to 'reverse engineer' products that are commercially marketed, reflecting a sense that maintaining this right permits more rapid scientific advance. Hence, it is possible that, even if they would otherwise be enforceable under contract law principles, these agreements are unenforceable because preempted by federal standards on intellectual property protection (or, in other legal systems, by a competition law provision). The leading recent Supreme Court example is Bonito Boats, Inc. v. Thunder Craft Boats, Inc. (489 US 141 (1989)), which struck down a state statute prohibiting the use of direct molding processes to copy boat hulls, on the theory that the state statute 'conflicts with "strong federal policy favoring free competition in ideas which do not merit patent protection"'. (The quotation is from an earlier case dealing with patent licences (Lear, Inc. v. Adkins; 395 US 653 (1969)).) There was also an early Plant Patent Act decision which regarded as an antitrust violation a contract between a breeder and its distributors that contained a number of restrictive provisions including one under which the original breeder sought to retain title to all sports deriving from the supplied material (Yoder Bros. Inc. v. California–Florida Plant Corp., 537 F.2d 1347 (1976)). Nevertheless, in 1996, a federal judge in the US Midwestern area upheld a somewhat parallel agreement governing use of a CD-ROM containing an uncopyrightable database (ProCD, Inc. v. Zeidenberg, 86 F.3d 1447 (7th Cir. 1996)). Much of the new case's logic could be applied by analogy to the seed labels – but will not necessarily be followed in other regions of the nation.

ENFORCEMENT

Enforcement of all of these rights is by private suit before a court (except for certain uses of intellectual property rights to prevent imports of a protected product or of the product of a protected process, in which case suit may be before the International Trade Commission under 19 USC §1337). The process is dependent on the initiative of the holder of the right, who generally has the burden of proof to demonstrate infringement, which, in the case of a patent, means showing that the allegedly infringing variety is within the scope of the claims of the patent. Although there is a presumption that the patent is valid, the defendant may attempt to show that the patent is invalid, as by showing that there was previous publication, that the invention was obvious, or that the patent disclosure was not enabling or did not reflect the patentee's best mode of performing the invention at the time of filing (35 USC §112). If the plaintiff succeeds, it can frequently obtain an injunction against use of the product (35 USC §283), in addition to damages, which are based on its actual market loss or on an estimate of a reasonable royalty (35 USC §284). In the case of trade secrecy, damages can also include a requirement that the defendant disgorge any profits gained from use of the secret.

The process can be very expensive, reaching in the USA about US$500,000 per side per claim litigated. This is a result of the legal fees and of the expenses spent in each side's effort to obtain information from the other. Expenses are especially high in the USA, because that nation still has a 'first to invent' system, implying that two firms, each seeking to demonstrate that it was the first to invent, will have to present evidence about the detailed history of the research process. Moreover, there may be extensive research through obscure journals in an effort to show that the invention was not novel. There may also be substantial expert testimony about the precise interpretation of the claims, and there may be a need to develop significant scientific evidence in order to demonstrate the similarity of two varieties.

The realities of contemporary litigation in this area are exemplified in Pioneer Hi-Bred International v. Holden Foundation Seeds, 35 F.3d 1226 (8th Cir. 1994). This was a trade secrecy suit, in which Pioneer claimed that Holden had used one of its inbred corn lines in the development of competing lines. The case was tried before a judge and the judge admitted evidence from isozyme electrophoresis, reverse phase HPLC and growout tests. These demonstrated substantial similarity between the Pioneer and the Holden lines. Holden was then unable to provide evidence persuading the court that it had developed the line independently in a way that did not infringe Pioneer's rights. It lost a judgment for over US$46 million.

Such litigation is rare, because it is so expensive, and there have been very few suits over specific lines. Among the important exceptions is a case holding that a patent under the Plant Patent Act can be infringed only by an asexually propagated product of the protected variety (Imazio Nursery, Inc. v. Dania Greenhouses, 69 FR2d. 1560 (CAFC, 1996)). At this time, firms appear to be using their litigation budget primarily for disputes over fundamental biotechnology patents, e.g. rights in various aspects of the use of *Bacillus thuringiensis* as in Plant Genetic Systems v. Mycogen Plant Science, Inc. 933 F.Supp 514 and 519 (MDNC, 1996), and Mycogen Plant Science, Inc. v. Monsanto Co., 1995 US Dist. LEXIS 20383 (SDCa, 1995), rather than for disputes over specific lines, and they have, of course, been seeking to avoid litigation through building portfolios of patents to be used defensively or for cross-licensing. It may therefore be some time before we have solid judicial answers to the uncertain issues discussed above.

REFERENCES

Anon. (1985) Plant life – Patentable Subject Matter. *Off. Gaz.* 160, 4 (8 October, 1985).

Baenziger, P.S., Kleese, R.A. and Barnes, R.F. (eds) (1993) *Intellectual Property Rights: Protection of Plant Materials.* Crop Science Society of America, Special Publication no. 21.

Barton, J. (1992) Biodiversity at Rio. *Biosciences* 42, 773 (November 1992).

Bruzzone, L. (1993) The Research Exemption: A Proposal. *American Intellectual Property Law Association Quarterly Journal* 21, 52.

Budapest Treaty (1977) Budapest Treaty of 28 April 1977 on the International Recognition of the Deposit of Micro-organisms for the Purpose of Patent Procedure.

Butler, L. and Marion, B. (1985) *The Impacts of Patent Protection on the U.S. Seed Industry and Public Plant Breeding.* North Central Regional Research Publication 304.

Eisenberg, R. (1989) Patents and the progress of science: exclusive rights and experimental use. *University of Chicago Law Revew* 56, 1017.

Food and Agriculture Organization of the United Nations (1983) *International Undertaking on Plant Genetic Resources.* Resolution 8/83 of the Twenty-second Session of the FAO Conference, 5–23 November 1983.

Hamilton, N. (1993a) Who owns dinner: evolving legal mechanisms for ownership of plant genetic resources. *Tulsa Law Journal* 28, 587.

Hamilton, N. (1993b) Legal issues in contract production of commodities: issues for farmers and their lawyers. Presentation at the American Agricultural Law Association, San Francisco, California, 12 November 1993.

Muirhead, Jr, R.C. and Shaver, D.L. (1985) Genetic factor capable of altering leaf number and distribution in maize. US Patent no. 4,513,532 (30 April, 1985).

Parr, P. (1993) Developments in agricultural biotechnology. *Wm. Mitchell Law Rev.* 19, 457.

Piper, T.E. (1997) Inbred maize line PHDG1. US Patent no. 5,602,318 (11 February 1997).

Roberts, T. (1996) Patenting plants around the world. *European Intellectual Property Review,* p. 531.

Seay, N. (1993) Intellectual Property Rights in Plants. In Baenziger, P.S., Kleese, R.A. and Barnes, R.F. (eds), *Intellectual Property Rights: Protection of Plant Materials.* Crop Science Society of America Special Publication no. 21, p. 61.

Stone, R. (1995) Sweeping patents put biotechnology companies on warpath. *Science* 268, 656 (5 May 1995).

Strissel, J., Pollak, G. and Kindiger, B. (1992) High lysine corn. US Patent no. 5,082,993 (21 January 1992).

Umbeck, P.F. (1992) Genetic engineering of cotton plants and lines. US Patent no. 5,159,135 (27 October 1992; reexamination granted 7 December 1994).

UPOV (International Union for the Protection of New Varieties of Plant) (1978) International Convention for the Protection of New Varieties of Plants of 2 December, 1961, as revised at Geneva on 10 November, 1972 and on 23 October, 1978.

UPOV (International Union for the Protection of New Varieties of Plant) (1991), International Convention for the Protection of New Varieties of Plants of 2 December, 1961, as revised at Geneva on 10 November, 1972, on 23 October, 1978, and on 19 March, 1991.

Zaitlin, M., Golemboski, D. and Lomonossoff, G. (1997) Induction of resistance to virus diseases by transformation of plants with a portion of a plant virus genome involving a read-through replicase gene. US Patent no. 5,596,132 (21 January, 1997).

Transferring Intellectual Properties

3

Frederic H. Erbisch and Andrew J. Fischer

Office of Intellectual Property, 238 Administration Building, Michigan State University, East Lansing, MI 48824-1046, USA

INTRODUCTION

The use of protected intellectual properties (IPs) is prohibited unless the owners allow others to use them. The owners can 'tie up' the IPs for the term of protection, but most protected IP has value to the inventor and/or to others. In general, the creator or discoverer of an IP or its owner will want to transfer the IP to gain either fame or financial rewards. If the value is primarily to the creator/discoverer, it is often kept by the creator/discoverer or freely shared with others. More commonly, the creator/discoverer will attempt to commercialize the IP; in essence, the creator/discoverer can trade the IP for money through the sale or licensing of the IP. Licensing is the most common method of transferring technology.

This chapter will focus on the various methods of transferring technologies, and in particular, will emphasize licensing of an invention. For clarity the word 'invention' will be used instead of 'intellectual property' and 'inventor' will be used in place of 'creator/discoverer' throughout this chapter. However, this does not mean the information only applies to inventions: it will apply to any IP. The examples will be based on experiences in the USA.

FREE, PUBLIC DISTRIBUTION OF INTELLECTUAL PROPERTY

Free, public distribution of IP is one method of distributing it. It rarely occurs in the biotechnology arena because inventors want to recoup costs associated with the invention, and to gain financially. Most inventors believe their inventions are valuable even though it appears they have little value. However, occasionally the inventor will give away an invention in exchange for another invention. The inventor may also receive good will or recognition.

For example, suppose an inventor develops a process (P_1) for creating substance S. P_1 is more time-consuming and more costly than other processes ($P_2, P_3, P_4 \dots$) used to create substance S, but is elegant in its methodology. The invention is patentable. Yet instead of patenting, the inventor decides to disclose the invention at the annual meeting of inventors because no one will ever use the invention commercially. By giving away the otherwise valueless invention, the inventor earns good will and praise from colleagues for the elegant method of producing substance S.

Besides good will, giving an IP away can be an excellent way to market it. One industry that relies on this method is the computer software industry. Some software companies give away a smaller, scaled-down version of their product to entice users into purchasing a license for a fully functional product. Other software inventors program the software to stop functioning after a certain date. In both cases, the creator/discoverer has control because they can program the software. This prevents the user from using the product in a manner contradictory to the owner's wishes. Free public distribution is used basically as a marketing tool to advertise the product in hopes of securing a commercial license. Biotechnology differs from software in that once information is given to the user or potential licensee, there is usually no way to restrict its use.

SALE OF INTELLECTUAL PROPERTY

Selling an invention is one method of commercializing it. Sale of an invention is called an assignment. Assignment of a patent occurs in either one of three ways; the owner of the patent can convey: (i) the whole patent, comprising the exclusive right to make, use or sell the patented invention throughout the USA; (ii) an undivided part or share of that exclusive right; or (iii) the exclusive right under the patent throughout a specified geographical location.

There are two primary problems associated with the sale of IP. The first is that the sale price cannot be determined. At what price should the inventor sell the property? This is a difficult question. The inventor's sale of IP happens only once. Therefore, the inventor must recoup, in the sale price, all monetary value in one transaction. Since most biotechnology inventions are not fully developed at the sale date, it is extremely difficult to put a monetary value on the invention. If the price is too high, no-one will purchase the invention. If the price is too low, the inventor loses money when the invention becomes commercially valuable. The second problem associated with the sale of IP is that the seller loses their rights to use the invention without permission from the new owner. Once the inventor sells the patent, *all* their interest in that invention is sold. Since, by definition, patent rights are the right to prevent all others from making, using or selling the invention, even the inventor will be prohibited from using the invention – even in their own research! In summary, it is uncommon for the inventor's research to start with a completely independent idea. Instead, the inventor usually bases current research on one or more of his/her previous inventions or ideas. If a sale of the invention were to occur at this midpoint, before an idea was completely developed, not only would the sale price be very difficult to determine, but the inventor would be prevented from utilizing his/her own research in any future inventions.

LICENSING THE INTELLECTUAL PROPERTY

To solve the sale price and future use problems associated with assignment or sale of a patent, the invention can be licensed. A license is a binding, revocable privilege to use the IP, for a fixed number of years, in a fixed territory in exchange for money or other compensation. It is a contractual relationship, and in the US, its enforcement is governed by contract law.

IP that can be licensed includes patents, trademarks, copyrights, trade secrets or other recognized forms of IP. Licensing has two distinct advantages over sale of the IP. The first is that the inventor retains ownership of the IP. Traditionally, ownership of any property carries with it certain rights. In the case of patents, this includes the right to forbid others from using, making or selling the invention. Trademarks, copyrights and the other forms of IP have their own rights associated with ownership. By retaining ownership, the inventor is assured of being able to protect those rights, and can sue those who infringe or use the patented rights without the owner's permission. Results of an infringement lawsuit may be huge. The inventor who successfully sues an infringer can get money (called *damages*) and/or a court order telling the infringer to stop making, using or selling the invention (this is called an *injunction*).

Ownership also carries with it certain implied rights. One such implied right in the USA is the right to 'shelve' the patent. (Note that some countries require an owner to exercise or use their patent or otherwise risk losing it.) Shelving means, figuratively, placing the patent on a shelf and doing nothing with it. For example, an inventor has a patent on P_1, the process for creating substance S. Process P_1 uses the raw material M to create substance S. The inventor also sells the raw material M, and is making lots of money on its sale. Next, the inventor invents and patents P_2, another process for making substance S. But process P_2 does not require the use of raw material M. To continue the sale of raw material M, the inventor shelves the P_2 patent.

The second advantage of licensing over sale of IP is that the inventor can license out the invention. The inventor can contract out some rights, and retain other rights. For example, the inventor can license out all rights to the invention with the exception of retaining the right to use the invention for non-commercial purposes. This retainer right by the inventor enables the inventor to continue his/her research, and is very common in licensing agreements.

An additional example of licensing a part of IP is the sublicense. The one who licenses the invention (licensee) is given the right to have others practise or do those things (use, make and/or sell) which are available to the licensee. For example, the inventor who patented process P_1 licenses the patent on process P_1 but retains the right to use the invention for non-commercial purposes. The licensee, who has been given the right to sublicense, licenses out (sublicenses) the patent to companies in Asia, Britain and Canada with restrictions that each can only sell, make or use the invention in their respective country.

IMPORTANT COMPONENTS OF A LICENSING AGREEMENT

A license agreement is a personal, revocable privilege that gives the licensee (usually the recipient of the technology) a right not to be sued by the licensor for using an invention. The license is primarily used for voluntary exchange of an invention for money or some other consideration. Because people have different understandings as to what is agreed in this transfer of technology, a written document is produced which exactly describes the obligations of both parties to the transfer transaction. This document is the license agreement. By putting into writing exactly what the parties intended, the license not only guides the parties to what they can and cannot do in the future, it also provides a dispute mechanism to which the parties can refer when misunderstandings or disagreements do occur. That is, the properly crafted license prevents disagreements but, if a misunderstanding does occur, it helps fashion a workable remedy.

Unfortunately, these disagreements occur often and are usually settled in court. In court, licenses are governed by contract law. Therefore many components of a license are dictated by what contract law requires. The remaining elements in the license are particular to the technology involved. (For example, although granting permission to reproduce a particular plant would be applicable in certain circumstances, it would be a meaningless provision if the technology were a mechanical device.)

While keeping these substantive requirements in mind, it is important that a license agreement is written clearly, avoiding ambiguity. Having said that, pinpointing every little detail and thinking through every possible contingency would not only take large amounts of time, but would produce a license document so large that it would be impractical to use. The key is to balance the need for certainty with the need for a practical workable document.

When deciding how much detail to put in the license, first keep in mind the value of the technology involved. The license for a product valued at US$1500 may have fewer pages and requirements than a ten year license valued at US$10,000,000. Other factors to consider are the parties' background, knowledge, industry practices or standards, and the desire for flexibility when an unforeseen circumstance occurs.

Also, try to avoid adding 'filler' to the license. That is, avoid legal terms that add ambiguity and uncertainty. Make each sentence clear, understandable and succinct. It is difficult for others to ascertain what the parties intended if they must re-read the sentence six or seven times or if one must consult an attorney every time the agreement is read.

Finally, label the parts of the license. Break the license down into sections or articles and give each an appropriate title. This allows one to refer to other sections of the license with pinpoint accuracy. Although there is no one license that will work in all situations, there are provisions that are common to most licenses. Most of these common requirements are dictated by contract law – that is they apply to all technology licenses, irrespective of the fact that the license is biotechnological or a mechanical device. A number of these 'common ' provisions are listed in Table 3.1. Each provision will be briefly described and the purpose of each reported. The basic licensing agreement used at Michigan State University (MSU) is included in the Appendix at the end of this chapter and will be refered to as the provisions are described. Seldom is the basic university license agreement used as

Table 3.1. Basic components of a license agreement.

1. The parties	6. Research support
2. Whereas clauses	7. Reporting requirements
3. Definitions	8. Diligence
4. Grant of license	9. Termination
5. Financial considerations	10. Liability/warranty
(a) initial payments	11. Use of names
(b) running royalties	12. Agreement governance
(c) minimum royalties	13. 'Boilerplate'

presented here; rather, it is modified for the particular invention being licensed and the potential licensee.

The following are brief reviews of some of the components of a license agreement.

1. The parties. Usually the parties of the license are named in the first paragraph of the license agreement. The MSU license lists the University as the licensor – that is, the party that is licensing – and the other party as the licensee, i.e. the one obtaining the right to use a patent/technology. Each party's full name and address should be included. If a party has more than one principal place of business, make a note of that in the licence. This prevents confusion between companies with the same name. When dealing with corporations in more than one country, always state the name of the country in the address.

After the names of the parties, a short-hand, capitalized notation is given in parenthesis. This name is used in the rest of the document so that the entire name need not be written each time. In the MSU license (see Appendix 1 at the end of this chapter) 'MSU' is the licensor and the other party is the licensee.

2. Whereas clauses. This portion of the license gives the basis for the agreement. These clauses list certain facts about the licensee, the technology and the licensor which simply state the position of the two parties to make the license arrangement possible.

3. Definitions. Definitions are critical in technical and scientific documents and especially in legal documents. Definitions are very important in the license agreement because many terms have more than one meaning. Remember it is important that there is no ambiguity in the license and that both parties understand the terms of the agreement.

In the MSU agreement (Appendix) definitions are found in Article I. Here definitions include licensed patent rights, products and net sales. Other definitions often added describe the 'field' of use for the inventions being licensed, i.e. the 'territory' in which the licensee can operate. Additional definitions may describe genes, plant types or varieties and other technology-specific terms used later in the license.

4. Grant of license. This is a very important part of the license. Through this provision the licensee is granted the right to manufacture, sell or use the invention (Article II). The licensee may be granted an exclusive license or a non-exclusive license. The exclusive license assures the licensee that the invention will not be licensed to any other party for commercial use. With a non-exclusive license the licensee may have competitors because the licensor can license the technology to another party or parties. The exclusive license can have variations too: the license can be exclusive for a geographic region rather than worldwide, or for a particular product rather than for all products which could be produced using the technology. The term of this license can be limited or can last for the life of the licensed patents and new patented improvements which are added to the license as required (Article IX).

If the license is exclusive and the licensor wants to continue to do research on the IP it is necessary to add a statement to the granting clause that the licensor reserves the right to continue to do non-commercial research and development (Article II).

5. Financial considerations. Usually the licensor does not grant the license without some financial consideration. There are three basic areas for financial consideration: initial payment, royalties and patent costs. The initial payment is made at the time the license is signed by both parties (Article III, Part 1). The amount is agreed through negotiation. The amount of payment depends upon the type of technology, the stage of development, the life of the patents and the company. If the initial payment is low then the royalty rate is generally higher than when the initial payment is large. Usually the royalty is based upon the sale of the product, actually the net selling price of the product (Article I, Part 4). The royalty rate is a percentage of the net selling price (Article III, Part 1). Rarely are royalties based upon licensee profits because of the difficulty of determining profit. To make sure the licensee does not shelve the licensed technology a provision for an annual minimum royalty payment is included in the license agreement (Article III, Part 4). Patent costs are very high, especially when foreign protection is also sought. The license provides that the licensee pays all these costs and in the case of foreign patents, the licensee is given the responsibility of deciding if foreign filing is to be done and in what countries to file (Article XI). In some instances, the licensee negotiates the right to deduct a portion of the patent costs from royalties.

6. Research support. In the case of university technologies few are completely developed and most need further research. The licensee is given an opportunity to have the inventor continue research on the invention. The actual research will be governed by a separate research agreement, but the fact the licensee will support research can be noted in the license agreement.

7. Reporting requirements. In order to ascertain the commercialization of the technology and the basis for royalty payment the licensee is required to submit required periodic reports. The royalty payment is due at the time the report is submitted (Article IV). The provision on diligence (Article VI) also has reporting requirements, but these reports are required only for a limited time and contain information of steps taken toward commercialization; these reports are very different from the required royalty-type reports.

8. Diligence. This provision is included in the license to assure the licensor that the licensee will move ahead commercially with the invention. The reporting requirements of this provision provide the licensor the satisfaction of knowing how the invention is being developed for commercialization (Article VI).

9. Termination. This provision provides a means for the licensee to terminate its relationship with the licensor, as well as for the licensor to terminate the arrangement. For the licensor to terminate and recover the technology the conditions must be such that commercialization of the licensed technology is in jeopardy. Without this provision the licensee could shelve, in some manner, the licensed technology and the licensor's technology would never be commercialized (Article VIII).

10. Liability/warranty. Once the licensee begins to make, sell and/or use the licensed technology, the licensor does not want to be responsible or liable for product so a provision provides that the licensee is responsible (Article XIII, Part 4). Generally, the invention licensed needs further development by licensee. Since the licensor has not taken the invention to the level of commercialization the licensor cannot warrant that the technology will be free of defects at this higher level of development (Article XIII, Part 10). While the licensor has used a patent attorney to draft and prosecute protection for the invention, and the US PTO has granted a patent, the licensor still cannot be sure that some company will not sue for infringement. Therefore, to protect itself the licensor includes a provision stating that it does not guarantee that the 'patent will be free of claims of infringement' (Article XIII, Part 3).

11. Use of names. One of the ways in which the licensor is able to control the licensee is by not allowing the licensee to use the licensor's name in advertizing. This prevents the licensee from using the licensor's name to endorse a product or imply that the licensor warrants or guarantees the product (Article XIII, Part 7).

12. Agreement governance. The licensor wants to have any legal actions taken care of near the licensor's facilities to minimize any legal costs. This provision of the agreement names the geographic area in which any legal action brought against the licensor by the licensee will be held. If the university licenses a technology to a company outside of the country, the provision will also state that the laws of the USA govern (Article XIII, Part 1).

13. 'Boilerplate'. Certain provisions included in a license agreement must be included because of contractual considerations. These provisions are rarely negotiated. Often these provisions are given the general name of 'boilerplate'. Both the licensee and the licensor know these provisions will be in the agreement and accept this condition.

SUMMARY

Protected IPs can provide the creator, discoverer or inventor with several options for taking it to the public. One method would be to make the IP freely available to anyone at no cost and under no obligations. Another way would be to sell the IP, but this means the originator of the IP loses all control of the property. The preferred way to transfer technology is through licensing because the originator of the property maintains control of it. The license agreement contains a number of provisions which the licensee is required to follow, all which are to the benefit, often financially, of the creator, discoverer or inventor.

APPENDIX 3.1: EXCLUSIVE LICENSE AGREEMENT

THIS AGREEMENT is made and entered into between Michigan State University, a not-for-profit corporation organized under the laws of the State of Michigan (hereinafter called "Licensor"), having its principal office at East Lansing, Michigan 48824, and _____ , a for-profit corporation organized under the laws of the State of _____ (hereinafter called "Licensee"), having its principal office at _____ .

WITNESSETH THAT:

WHEREAS, Licensor has the right to grant licenses under the Licensed Patent Rights (as hereinafter defined), and wishes to have the inventions covered by the Licensed Patent Rights utilized in the public interest; and

WHEREAS, Licensee wishes to obtain a license under the Licensed Patent Rights upon the terms and conditions hereinafter set forth:

NOW, THEREFORE, in consideration of the premises and the faithful performance of the covenants herein contained it is agreed as follows:

Article I: Definitions
For the purpose of this Agreement, the following definitions shall apply:
1. "Licensed Patent Rights" means:
    (a) U.S. patent application Serial No. _____ filed _____ by _____ .
    (b) Any and all Improvements developed by Licensor, whether patentable or not, relating to the Licensed Patent Rights.
    (c) Any and all patents which may issue on patent applications to be filed on Licensed Patent Rights and improvements thereof developed by Licensor and any and all divisions, continuations, reissues and extensions of such patents or applications, and including all United States and foreign counterpart applications and patents.
2. "Products" mans any materials, compositions, techniques, devices, methods or inventions relating to or based on the Licensed Patent Rights, developed on the date of this Agreement or in the future.
3. "Net Sales" means Gross sales, FOB Place of Manufacture of Products, less sales and/or use taxes, third party commissions, discounts, customs duties, and shipping.

Article II: Grant of Exclusive License
Licensor hereby grants to Licensee the exclusive worldwide license with the right to sublicense others, to make, have made, use, and sell the Products described in the Licensed Patent Rights. Licensor reserves the right to continue to use the Licensed Patent Rights in its non-commercial research programs.

Article III: License Payments
1. Initial Payment and Royalty Rate.
    For the license herein granted:
    (a) Licensee agrees to pay a sign-up fee of _____. Payment shall be made as follows: _____ .

(b) Licensee shall pay an earned royalty of _____ of Licensee's and its sublicensee's Net Sales of Products or fifty percent (50%) of the sublicensing receipt whichever is greater.

(c) No royalties shall be due for any product not directly a part of the Products.

2. *Sublicensees*

The granting and terms of all sublicenses is entirely at Licensee's discretion provided that all sublicenses shall be subject to the terms and conditions of this Agreement.

3. *Minimum royalties*

Licensee will pay Licensor, when submitting their Royalty Report for the first quarter of the following annual period, the difference between the specified minimum royalty and the earned royalties for the annual period, if either:

(a) the earned royalties accrued and paid hereunder by Licensee, for a period beginning on _____ and ending on _____, are less than _____ dollars _____; or

(b) such earned royalties for any thereafter annual period, through the term of this Agreement, are less than _____ dollars _____ .

4. *When a sale is made*

A sale of Licensed Patent Rights shall be regarded as being made upon payment for products made using Licensed Patent Rights. Royalties paid to Licensor by Licensee where there is a return or non-acceptance by the customer and for which a refund is given to the customer may be credited in an appropriate amount against future royalties payable hereunder.

5. *Payments in US*

All sums payable by Licensee hereunder shall be paid to Licensor in the United States and in the currency of the United States.

6. *Interest*

In the event any royalties are not paid as specified herein, then a compound interest of eighteen percent (18%) shall be due in addition to the royalties accrued for the period of default.

Article IV: Reports, Books and Records

1. *Reports*

Within thirty (30) days after the end of the calendar quarter annual period during which this Agreement shall be executed and delivered and within thirty (30) days after the end of each following quarter annual period. Licensee shall make a written report to Licensor setting forth the total Net Sales of Licensed Patent Rights sold, or used by Licensee and total sublicensing receipts during the quarter annual period. If there are no Net Sales or sublicensing receipts, a statement to that effect shall be made by Licensee to Licensor. At the time each report is made, Licensee shall pay to Licensor the royalties or other payments shown by such report to be payable hereunder.

2. *Books and Records*

Licensee shall keep Books and Records in such reasonable detail as will permit the Reports provided for in Section 1 hereof to be made and the Royalties payable by Licensee hereunder to be determined. Licensee further agrees to permit such Books and Records to be inspected and audited from time to time (but not more often than once semi-annually) during reasonable business hours by a representative or representatives of Licensor to the extent necessary to verity the Reports provided for in Section 1 hereof; provided, however, that such representative or representatives shall indicate to Licensor only whether the Reports and Royalty paid are correct, if not, the reasons why not.

Article V: Marking

Licensee agrees to mark or have marked all products made, sold, or used by it or its sublicensees under the Licensed Patent Rights, if and to the extent such markings shall be practical, with such patent markings as shall be desirable or required by applicable Patent Laws.

Article VI: Diligence

1. Licensee shall use its best efforts to bring Licensed Patent Rights to market through a thorough, vigorous and diligent program and to continue active, diligent marketing efforts throughout the life of this Agreement.
2. In addition, Licensee shall adhere to the following milestones:
    (a) Licensee shall deliver to Licensor on or before _____, 199_, a business plan for the development of Licensed Patent Rights, which includes number and kind of personnel involved, time budgeted and planned for each phase of development of Licensed Patent Rights. Quarterly reports describing progress toward meeting the objectives of the business plan shall be provided through _____, 200_.
    (b) Licensee shall develop a commercializable Product developed from Licensed Patent Rights on or before _____ 200_ and permit an in-house inspection of Licensee's facilities by Licensor on an annual basis beginning _____, 199_ .
3. Licensee's failure to perform in accordance with either paragraph 1. or 2. of this ARTICLE VI shall be grounds for Licensor to terminate this Agreement.

Article VII: Irrevocable Judgment With Respect to Validity of Patents

If a judgment or decree shall be entered in any proceeding in which the validity or infringement of any claim of any patent under which the license is granted hereunder shall be in issue, which judgment or decree shall become not further reviewable through the exhaustion of all permissible applications for rehearing or review by a superior tribunal, or through the expiration of the time permitted for such application (such a judgment or decree being hereinafter referred to as an "Irrevocable Judgment"), the construction placed on any such claim by such Irrevocable Judgment shall thereafter be followed not only as to such claim, but also as to all claims to which such construction applies, with respect to acts occurring thereafter and if an Irrevocable Judgment shall hold any claim invalid, Licensee shall be relieved thereafter from including in its reports hereunder that portion of the royalties due under Article III payable only because of such claim or any broader claim to which such Irrevocable Judgment shall be applicable, and from the performance of any other acts required by this Agreement only because of any such claims.

Article VIII: Termination or Conversion to Non-Exclusive License
1. *Termination by Licensee*
    (a) *Option of Licensee to terminate.* Licensee may terminate the license granted by this Agreement, provided Licensee shall not be in default hereunder, by giving Licensor ninety (90) days notice of its intention to do so. If such notice shall be given, then upon the expiration of such ninety (90) days the termination shall become effective; but such termination shall not operate to relieve Licensee from its obligation to pay Royalties or to satisfy any other obligations, accrued hereunder prior to the date of such termination.
    (b) *Option of Licensee to convert to non-exclusive license.* Licensee shall have the right to convert this license to a non-exclusive license at the same royalty rate as for the exclusive license, without right to sublicense and minimum Royalties under Article III (3) shall not be due thereafter. Upon termination of this Agreement, all of the Licensed Patent Rights shall be returned to Licensor. In the event of termination of the Agreement by Licensee or said conversion of the Agreement by Licensee, Licensee shall grant to Licensor a non-exclusive, royalty-free license, with right to sublicense, to manufacture, use and sell Improvements including all know-how to Licensed Patent Rights made by Licensee during the period of this Agreement prior to the termination or conversion, to the extent that such Improvements are dominated by or derived from the Licensed Patent Rights.
2. *Termination by Licensor*
    (a) *Default by Licensee.* Licensor may, at its option, terminate this Agreement by written notice to Licensee, if Licensee shall default in:
        (1) The payment of any Royalties required to be paid by Licensee to Licensor hereunder or in the making of any reports required hereunder and such default shall continue for a period of thirty (30) days after Licensor shall have given to Licensee a written notice of such default; or
        (2) The performance of any other material obligation contained in this Agreement on the part of Licensee to be performed and such default shall continue for a period of thirty (30) days after Licensor shall have given to Licensee written notice of such default.
    (b) *Bankruptcy. etc.* Licensor shall have the right, by written notice to Licensee, to terminate this Agreement at any time upon or after;
        (1) An adjudication that Licensee is bankrupt or insolvent;
        (2) The filing by Licensee of a Petition of Bankruptcy, or a Petition or Answer seeking reorganization, readjustment or rearrangement of its business or affairs under any law or governmental regulation relating to bankruptcy or insolvency;
        (3) The appointment of a receiver of the business or for all or substantially all of the property of Licensee;
        (4) The making by Licensee of an assignment or an attempted assignment for the benefit of its creditors; or
        (5) The institution by Licensee of any proceedings for the liquidation or winding up of its business or affairs.
    (c) *Effect of termination.* Termination of this Agreement shall not in any

way operate to impair or destroy any of Licensee's or Licensor's rights or remedies, either at law or in equity, or to relieve Licensee of any of its obligations to pay Royalties or to comply with any other of the obligations hereunder, accrued prior to the effective date of termination.

(d) *Effect of delay etc.* Failure or delay by Licensor to exercise its rights of termination hereunder by reason of any default by Licensee in carrying out any obligation imposed upon it by this Agreement shall not operate to prejudice Licensor's right of termination for any other subsequent default by Licensee.

Article IX: Term

Unless previously terminated as hereinbefore provided, the term of this Agreement shall be from and after the date hereof until the expiration of the last to expire of the licensed issued patents or patents to issue under the Licensed Patent Rights under Article I. Licensee shall not be required to pay royalties due only by reason of its use, sale, licensing, or sublicensing under issued patents licensed by this Agreement that have expired or been held to be invalid by an Irrevocable Judgment, where there are no other of such issued patents valid and unexpired covering the Licensee's use, sale, licensing, or sublicensing; provided, however, that such non-payment of royalties shall not extend to royalty payments already made to Licensor more than six (6) months prior to Licensee's discovery of expiration or an Irrevocable Judgment.

Article X: Patent Litigation

1. *Initiation*

   In the event that Licensor advises Licensee in writing of a substantial infringement of the patents included in the Licensed Patent Rights, Licensee may, but is not obligated to, bring suit or suits through attorneys of Licensee's selection with respect to such infringement. In the event Licensee fails to defend any declaratory judgment action brought against any patent or patents of the Licensed Patent Rights, Licensor on written notice to Licensee may terminate the license as to the particular Patent or Patents involved in such declaratory judgment action.

2. *Expenses and proceeds of litigation*

   Where a suit or suits have been brought by Licensee, Licensee shall maintain the litigation at its own expense and shall keep any judgments and awards arising from these suits excepting that portion of the judgments attributable to Royalties from the infringer shall be divided equally between Licensor and Licensee after deducting any and all expenses of such suits; provided, however, Licensor shall not be entitled to receive more under this provision than it would have received if the infringer had been licensed by Licensee.

3. *Licensor's right to sue*

   If Licensee shall fail to commence suit on an infringement hereunder within one (1) year after receipt of Licensor's written request to do so, Licensor in protection of its reversionary rights shall have the right to bring and prosecute such suits at its cost and expense through attorneys of its selection, in its own name, and all sums received or recovered by Licensor in or by reason of such suits shall be retained by Licensor; provided, however, no more than one lawsuit at a time shall commence in any such country.

ARTICLE XI: Patent Filings and Prosecuting

1. Licensee shall pay future costs of the prosecution of the patent application presently pending in the United States Patent and Trademark Office as set forth in Article I (2) which are reasonably necessary to obtain a patent. Furthermore, Licensee will pay for the costs of filing, prosecuting and maintaining foreign counterpart applications to such pending US Patent Applications, such foreign applications to be filed within ten (10) months prior to the filing date of the corresponding US Patent Application.

2. Improvements by the Inventors shall be owned by Licensor. Licensee shall pay fixture costs of preparation, filing, prosecuting and maintenance of patents and applications on patentable Improvements made by the Inventors; however, in the event that Licensee refuses to file patent applications on such patentable Improvements in the US and selected foreign countries when requested by Licensor, the rights to such patentable Improvements for said countries shall be returned to Licensor.

3. The patent preparation, prosecution and maintenance of patent applications and patents undertaken at Licensee's cost shall be performed by patent attorneys selected by Licensor; and due diligence and care shall be used in preparing, filing, prosecuting, and maintaining such applications on patentable subject matter. Both parties shall review and approve any and all patent related documents.

4. Licensee shall have the right to, on thirty (30) days written notice to Licensor, discontinue payment of its share of the prosecution and/or maintenance costs of any of said patents and/or patent applications. Upon receipt of such written notice, Licensor shall have the right to continue such prosecution and/or maintenance in its own name at its own expense in which event the license shall be automatically terminated as to the subject matter claimed in said patents and/or applications.

5. Notwithstanding the aforegoing Paragraph of this Article XI, Licensee's obligations under such Paragraphs shall continue only so long as Licensee continues to have an exclusive license under the Licensed Patent Rights and, in the event of conversion of the license to non-exclusive in accordance with Article VIII, Paragraph 1, after the date of such conversion:

   (a) The costs of such thereafter preparation, filing, prosecuting and maintaining of said licensed patents and patent applications shall be the responsibility of Licensor, provided such payments are at the sole discretion of the Licensor; and

   (b) Licensee shall have a non-exclusive license without right to sublicense under those of such patents and applications under which Licensee had an exclusive license prior to the conversion.

Article XII: Notices, Assignees

1. *Notices*
   Notices and payments required hereunder shall be deemed properly given if duly sent by United States First Class Mail and addressed to the parties at

the addresses set forth above. The parties hereto will keep each other advised of address changes.

2. *Assignees, etc.*

This Agreement shall be binding upon and shall inure to the benefit of the assigns of Licensor and upon and to the benefit of Licensee and the successors of the entire business of Licensee, but neither this Agreement nor any of the benefits thereof nor any rights thereunder shall, directly or indirectly, without the prior written consent of Licensor, be assigned, divided, or shared by Licensee to or with any other party or parties (except a successor of the entire business of Licensee).

Article XIII: Miscellaneous

1. *Law of Michigan governs*

This Agreement is executed and delivered in Michigan and shall be construed in accordance with the laws of that State.

2. *No other understanding*

This Agreement sets forth the entire agreement and understanding between the parties as to the subject matter thereof and merges all prior discussions between them.

3. *No representations or warranties regarding patents of third parties*

No representation or warranty is made by Licensor that the Licensed Patent Rights manufactured, used, or sold under the Exclusive License granted herein is or will be free of claims of infringement of patent rights of any other person or persons. The Licensor warrants that it has title to the Licensed Patent Rights from the Inventors.

4. *Indemnity*

Licensee shall indemnify, hold harmless, and defend Licensor and its trustees, officers, employees, students and agents against any and all allegations and actions for death, illness, personal injury, property damage, and improper business practices arising out of the use of the Licensed Patent Rights.

5. *Insurance*

During the term of this Agreement, Licensee shall, where commercially viable, use its best efforts to keep and maintain in full force and effect the following insurance coverage:

  (a) Comprehensive General Liabllity with limits of no less than One Million Dollars ($1,000,000.00) each occurrence and annual aggregate for bodily injury and property damage.

  (b) Professional Liability in the minimum amount of One Million Dollars ($1,000,000.00) each occurrence and annual aggregate.

  (c) Workers' Compensation consistent with statutory requirements.

  (d) General Property Damage Insurance with limits of no less than One Hundred Thousand Dollars ($100,000.00) per incident.

Such insurance shall be carried with companies rated A or better by A.M. Best. Certificates of Insurance shall be provided to Licensor upon request by Licensor. Before Licensee makes any change in any such policy or its coverage, including without limitation the termination thereof, Licensee shall notify Licensor as soon as possible. Any such change under the control of Licensee that would adversely affect the protection of Licensor by reducing the coverage available to Licensor below any level specified in this Article XIII, Paragraph 5. shall constitute a material breach of the Agreement by Licensee.

6. *Headings, etc.*
The titles or headings, articles, sections or paragraphs set forth in this Agreement have been inserted merely to facilitate reference and shall have no bearing upon the interpretation of any of the provisions of this Agreement.

7. *Advertizing*
Licensee agrees that Licensee may not use in any way the name of Licensor or any logotypes or symbols associated with Licensor or the names of any researchers without the express written permission of Licensor.

8. *Counterparts*
This Agreement may be executed in any number of counterparts, any one of which shall be deemed to be the original without the production of others.

9. *Confidentiality*
The parties agree to maintain discussions and proprietary information revealed pursuant to this Agreement in confidence, to disclose them only to persons within their respective companies having a need to know, and to furnish assurances to the other party that such persons understand this duty of confidentiality.

10. *Disclaimer of warranty*
Licensed Patent Rights is experimental in nature and it is provided WITH-OUT WARRANTY OR REPRESENTATIONS OF ANY SORT, EXPRESS OR IMPLIED, INCLUDING WITHOUT LIMITATION WARRANTIES OF MER-CHANTABILITY AND FITNESS FOR A PARTICULAR PURPOSE OR NON-INFRINGEMENT. Licensor makes no representation and provides no warranty that the use of the Licensed Patent Rights will not infringe any patent or proprietary rights of third parties.

IN WITNESS WHEREOF, the parties hereto have caused this Agreement to be executed by their duly authorized representatives.

The effective date of this Agreement is _____ , 19__ .

AGREED TO AND ACCEPTED: MICHIGAN STATE UNIVERSITY

_____ By:_____

Date of signature Name:_____

Title:_____

AGREED TO AND ACCEPTED: COMPANY NAME

_____ By:_____

Date of signature Name:_____

Title:_____

Capacity Building in Intellectual Property Management in Agricultural Biotechnology

4

Karim M. Maredia[1] and Frederic H. Erbisch[2]

[1]Institute of International Agriculture, 416 Plant and Soil Science Building, Michigan State University, East Lansing, MI 48824-1325 and [2]Office of Intellectual Property, 238 Administration Building, Michigan State University, East Lansing, MI 48824-1046, USA

INTRODUCTION

Capacity building is the strengthening and/or development of human resources and their institutional support structures. In agriculture, biotechnology is currently applied to improve agricultural productivity in order to feed growing populations in an environmentally friendly manner. Biotechnology encompasses many factors, not just research, but includes policy, networking and management. In order to utilize properly the new and emerging tools of biotechnology, nations must take an integrated approach and build capacity in all these areas. Developing capacity in policy areas includes fostering experience and expertise in intellectual property rights (IPR), biosafety and commercial linkages. This chapter focuses on capacity building in IPR, an important policy element for the proper use, access and exchange of new and emerging biotechnologies.

Large investments in new biotechnologies have been made by the private sector in developed countries. While many of these promising new biotechnologies reside in industrial countries, they may offer new ways in which developing countries could enhance their agricultural productivity. These technologies are often proprietary in nature and must be managed in a different way from non-proprietary technologies. In the past, technologies developed by the public sector were freely exchanged, particularly agriculture-related technologies such as crop varieties and germplasm.

The recent changes in the General Agreement on Tariffs and Trade (GATT) now require members of the World Trade Organization (WTO)

to respect each other's IPR. In the past, public sector research in most developing countries was predominantly supported by their governments which mandated that public sector institutions freely serve society. Therefore, technologies generated by the public sector were freely exchanged for research and development purposes without entering into any kind of commercial agreements. With the advent of biotechnology, this trend is now changing.

CAPACITY BUILDING IN INTELLECTUAL PROPERTY RIGHTS

As the global community increasingly attempts to privatize the agricultural sector, and access new and emerging technologies, many national governmental policies are changing to address IPR issues. In order to build a sound IPR framework, nations must address the issue of capacity building at both the national and institutional level. The following sections discuss capacity building at both national and institutional levels using the Agricultural Biotechnology for Sustainable Productivity (ABSP) project as a case study where appropriate.

The ABSP project started in September 1991 as a six year cooperative agreement between the United States Agency for International Development (USAID) and Michigan State University (MSU). The ABSP project is a consortium of public and private sector institutions in the USA and developing countries and represents an integrated approach to agricultural biotechnology research and development programs. It establishes links between a developing country's public and private sectors and the USA's public and private sectors (Maredia and Dodds, 1994). The ABSP project is currently assisting Egypt, Kenya, Indonesia, Costa Rica and Morocco in the use and management of biotechnologies. For example, one area on which the project is focusing is building capacity in IPR and technology transfer.

The integrated approach to IPR capacity building includes the following areas: (i) awareness creation; (ii) human resource development; (iii) institutional development; and (iv) information access. These areas are discussed in the following sections.

Awareness creation

Proper awareness of biotechnology-related issues must be created nationally. Since the IPR issues are closely tied in with the use and management of biotechnology, nations must create proper awareness among the general public, policy makers, scientists and administrators. It is very important that several approaches are used to create this awareness and many international programs are involved in this process. For example, the ABSP project is assisting developing countries on IPR issues through seminars, workshops, one-on-one consultations and information delivery. The project has organized several international and national workshops on various aspects of IPR, including an IPR workshop in Cairo, Egypt, in January 1994 (Bedford and Maredia, 1994). Over 150 scientists and senior administrators from various public and private institutions in Egypt attended this workshop. Also, in March 1996, the project organized a Plant Variety Protection (PVP) Workshop in Morocco attended by over 250 participants (Ives, 1997).

Since biotechnology is a relatively new field, most developing countries do not have appropriate national or institutional IPR policies. Scientists working within public sector institutions in developing countries are not fully aware of the importance of IPR issues and the access and exchange of proprietary technologies.

In the USA, associations such as the Association of University Technology Managers (AUTM), the Biotechnology Industry Organization (BIO), the American Seed Trade Association (ASTA), various federal agencies, and university intellectual property (IP) offices play an important role in creating general awareness of biotechnology-related IPR issues. Appendix 4.1 lists selected organizations and institutions that provide IPR-related information, and offer training and capacity building opportunities.

Human resource development

Human resource development addressing national policy issues
Since most biotechnologies are proprietary in nature, appropriate national policies related to IPR must be in place to access these technologies. When the ABSP project was initiated in 1991, none of the collaborating countries had IPR policies relating to agricultural biotechnology. Therefore, the project initially assisted in the area of appropriate IPR policy formulation. This formulation may require development, changes, revisions and/or amendments in legislation. To make these changes or revisions, policy makers must receive information regarding national and international biotechnology related IP management issues.

To meet these educational needs, the ABSP project, under the leadership of Professor John Barton, organized an internship program in IPR at Stanford Law School in 1993 (Barton and Bedford, 1993). The internship's goal was to educate both the policy makers and scientists in various agricultural biotechnology-related IPR issues. Subsequently, the ABSP project organized other IPR workshops and provided one-on-one consultations to help build a country's national IPR capacity. For example, in March 1995, the ABSP project organized a PVP and Patents Workshop in Jakarta, Indonesia and assisted in drafting a new PVP law for Indonesia. This PVP law is currently pending the approval of the parliament. In Egypt, ABSP-trained personnel facilitated the inclusion of plant and food products into existing IPR legislation. Therefore, the impact of ABSP human resources development activities has been tremendous.

In addition to scientists and policy makers, patent examiners need to be trained to examine biotechnology-related patent applications efficiently. IPR legislation in many nations now allows patenting of agricultural biotechnology products. However, the patent examiners who examine these applications for patentability do not have an adequate background and knowledge of biotechnology. The US Patent and Trademark Office (US PTO) in Washington DC conducts patent application and examination training. Utilizing this resource, ABSP trained two patent examiners and one scientist from Indonesia. In its capacity building activities, the ABSP project has always stressed the importance of developing linkages between legal and scientific personnel.

At the national level, capacity building is very important. Without proper awareness and education, policy makers and senior administrators who are responsible for formulating new policies cannot develop an appropriate IPR framework. Without an appropriate framework, scientists will not be able to access foreign technologies. Without access to new technology, increased agricultural production, necessary to feed a growing population, will probably not occur. The appropriate national IPR framework also allows countries and local scientists to protect their own inventions, and protect their genetic and biological resources.

Human resource development addressing institutional policy issues
The previous section discussed the IPR policies and capacity building issues at the national level. The real challenge is implementing and enforcing broad national IPR policies at the institutional level. In other words, national IPR policy issues must be institutionalized. Trained human resources must be available at the institutional level to help design and implement the agricultural biotechnology-related IPR policies.

IPR capacity building at the institutional level can have multiple benefits. Some benefits include assisting in the proper implementation of national policies and allowing institutions to protect, exchange and commercialize their own technologies, potentially generating revenues for further research and development. Globally, there is a steady decline in the financial support that governments provide to public institutions. In this declining environment, government policies around the world are shifting towards commercialization of agricultural research to sustain public institutions. Policy makers and senior administrators are re-thinking the ways in which technology from the public sector is handled and exchanged.

The capacity building at the institutional level needs to be addressed at three levels. First, IP protection needs to be institutionalized in relation to national policies. Second, persons responsible for the day-to-day handling of IPs should be well trained. Finally, scientists must be educated in the handling and management of IPs.

Currently, IP management lacks well trained people in developing countries. In order for institutions to function efficiently, human resources are essential. To help build capacity in IPR and technology transfer, the ABSP project, in cooperation with the Office of Intellectual Property at MSU, organized a two-week internship program in February 1996 (Maredia *et al.*, 1996). The internship program was aimed at providing hands-on experience with IP management on a day-to-day basis. The program also covered how technologies are transferred from the public to the private sector. The emphasis in the internship program was placed on the ideas, concepts and processes used in the handling, transfer and management of intellectual properties by various US institutions. The participants' goal was to learn and become familiar with these ideas and return to their countries with this knowledge. Eleven international participants from eight different countries in Africa, Asia and Latin America participated in the internship program. Due to the success of this program, planning is under way to conduct the internship program again in 1997.

Institutional development

Development of IP management focal points
As discussed earlier, IP management lacks focal points in most developing countries. This requires establishing points of contact within public and private institutions that deal with IP management. Public institutions in the USA and other developed countries have gone through similar experiences and adjusted their policies and institutional framework accordingly. For example, in the USA, the passage of the Bayh–Dole Act in the early 1980s provided the basis for public sector IP protection and technology transfer practices at universities. Taking advantage of these changes, public sector institutions reacted quickly and strengthened their IP protection and technology transfer framework through the creation of formal offices of intellectual property and/or technology transfer. MSU, for example, established its Office of Intellectual Property in 1992. This office plays an active role in the day-to-day management of intellectual properties. Thailand has recently established a similar focal point at Chulalongkorn University (L. Tanasugarn, California, 1997, personal communication).

Potential role of intellectual property/technology transfer offices
IPR and technology transfer programs (referred to hereafter as technology transfer office, or TTO) can play multiple roles in any institutions involved in research and development. These multiple roles are described below.

EDUCATION AND AWARENESS. Arguably, the most important role of the TTO is creating awareness and educating scientists in how to handle new inventions and discoveries. The TTO can conduct educational programs to make scientists aware of proper handling of inventions, including proper record-keeping, use of confidential disclosures, publication guidelines and the development of proper agreements. The office can also conduct on-campus seminars and training programs, and develop informational material.

CREATION AND MODIFICATION OF INSTITUTIONAL POLICIES RELATED TO INTELLECTUAL PROPERTY PROTECTION AND TECHNOLOGY TRANSFER. The TTO can help develop and enforce institutional policies dealing with inventions and discoveries. The office also ensures that the requirements of federally funded projects are met.

PROTECTION OF INTELLECTUAL PROPERTY. TTOs play important roles when reviewing inventions and determining their patentability and commercial potential. For example, faculty members disclose their inventions to the TTO, which then works with appropriate legal and business personnel evaluating new discoveries and inventions. During this process, if found useful, the TTO can also help protect and commercialize these new inventions. The TTO can also play a role in the event of infringement or litigation of IPs.

GENERATION OF NEW REVENUES THROUGH LICENSING OF INTELLECTUAL PROPERTIES. The TTO helps assess the commercial potential of IPs and markets these technologies through licensing, generating revenues. Licensing of technology involves promotion, marketing, negotiation, implementation and execution of the actual license agreement, including collecting royalty payments. The office can play a key role in all these aspects.

NETWORKING. Networking is also an important role of the TTO. For example, if the office maintains a database of all new technologies with commercial potential, this information can be shared with potential commercial partners. Additional benefits include networking with other technology transfer associations and combining different technologies into technology packages. This package may create more value than a single technology alone and enable execution of a licensing agreement.

CREATION OF NEW START-UP COMPANIES. The TTO, through links with venture capital firms, can help establish new start-up companies using its own technologies by putting inventors, investors, attorneys and business managers together. Also, the office can help educate researchers in what steps are required to start a new company.

SERVICE TO SOCIETY. Because society supports many public institutions indirectly through the payment of state and federal taxes, sharing the benefits of new technologies is a service the TTO office is obliged to provide. For example, transferring a new food crop variety to farmers through links with the private sector may increase food production, leading to lower prices for food. Additionally, revenues generated through the licensing of technologies can support continued research and development of new technologies.

Establishing an intellectual property/technology transfer office
Establishing a TTO is not an easy task. Issues that need to be addressed include institutional support for office operations, size of the office and the role the office will play. Continued financial support is critical for operating and managing the office. Experience in the USA and many other countries shows that, initially, these offices are not financially self-supporting and that the institution should provide operating funds until they become self-sufficient.

The size of the office includes the number of staff, the diversity and qualifications of personnel, and the size and location of the physical facility. In the USA, there are both big and small offices with small offices utilizing outside expertise as consultants on a need-based basis. Large offices may have specialists in various disciplines that carry the technology from its first disclosure through to commercialization. MSU, for example, has a relatively small office with four permanent staff members, and it uses outside expertise, especially patent attorneys and business personnel. On the other hand, the University of Michigan has an office with more than ten professionals covering nearly all aspects from invention disclosure to commercialization.

Regardless of office size, successful technology transfer requires the involvement of and/or interaction among technical, legal, business and financial personnel. The leader, director or coordinator of the office must establish a team framework and business plan to foster working relations and interactions among all these groups. Technical personnel may include broad areas of agriculture, basic sciences, engineering, medicine, etc. Legal personnel may include legal counsel, patent attorneys and infringement litigators. Business personnel include licensing and marketing individuals, with professional business liaisons. Financial personnel may include experts in venture capital generation.

In the USA, almost every public and private institution has some form of IP protection and technology transfer operation. Many different names are given to these operations, including 'Office of Intellectual Property' (MSU), 'Technology Management Office' (University of Michigan), 'Office for Technology and Trademark Licensing' (Harvard University) and 'Technology Licensing Office' (Massachusetts Institute of Technology, or MIT). Regardless of the name, their roles are essentially the same.

Day-to-day function and operation of office
Operation of TTOs varies depending on the size of the office, qualifications and roles of the personnel, the breadth of the office's mission and the size and location of the institution. The day-to-day operation would include many of the following elements:

- interactions with faculty, academic units, legal and business personnel;
- handling of inventions/disclosures and evaluation;
- protection of useful inventions and discoveries; material transfer agreements (MTAs), patents, copyright, trademark, trade secret, plant variety certification;˘
- maintenance of patents; foreign filing;
- licensing and marketing of technologies;
- new business development including start-ups;
- design and implementation of institutional policies relating to the handling and management of intellectual properties;
- educational seminars, training courses and informational materials for education and awareness purposes;
- networking with professional technology transfer organizations.

Day-to-day operation of the office of intellectual property at Michigan State University
MSU's Office of Intellectual Property (OIP) was established in 1992. The office handles intellectual properties and inventions developed or created by MSU faculty, staff and students. The office is under the supervision of the Vice-President for Research and Graduate Studies and supported, in part, by the MSU Foundation, a non-profit entity. Presently, the OIP has four staff members including the director, two licensing associates and one administrative assistant.

According to MSU's policy, any inventions developed using MSU facilities, funds or funds under the university's control are the property of MSU. These inventions are reported to the OIP. After the OIP has reviewed the disclosure with the inventor(s), a patent attorney is then consulted to review the invention for patentability. The patentability report is reviewed together by the OIP and the inventor. If the invention is patentable, the potential commercial value of the invention is discussed.

If the invention appears to have commercial value, the attorney is instructed to prepare a patent application and the OIP begins searching for an industrial partner or licensee. If all is successful, the industrial partner has a successful product, MSU receives royalties and society benefits. Royalties are shared by the inventor, the inventor's academic department and MSU.

Under the ABSP project, MSU's OIP has assisted developing counties to establish MTAs and research agreements with both public and private sectors in the USA. Additionally, the OIP has contracted with legal experts to ensure that developing country partners' interests are represented in negotiations with the US public and private sectors.

The OIP serves as the negotiating and licensing body for the university. Over 400 inventions were received by the OIP since 1992. Many are in various stages of patenting and licensing, with over 50 license agreements. It is too early to expect much royalty income from these newly licensed inventions, but MSU received approximately US$14 million in royalties during 1994, mainly from earlier licenses. The bulk of the royalties were earned through one license on a cancer treatment.

The OIP also plays an active role in marketing MSU technologies, implementing a system for non-proprietary description of technologies. The OIP is recognized in the marketplace throughout the US. Interactions with the marketplace are facilitated by utilizing several mechanisms. These include active participation in technology transfer shows and meetings, computer networking through national and international database listings, and development of the OIP's own World Wide Web page at http://web.miep.org/oip/oip.html. This site contains information on over 50 non-confidential disclosures available for licensing.

The office is also involved in entrepreneurial activities. Since its inception, the OIP has been involved in establishing five new companies based on MSU technologies. The OIP also reviews IP agreements for the MSU Contracts and Grants Office when new proposals are submitted and joint venture agreements are issued.

The OIP is also involved in education. The office plays an important role by educating the MSU faculty, staff and international visiting scholars in IP transfer and management issues. Periodically, the OIP, in conjunction with other departments, holds joint seminars which provide information and training in IP handling. Finally, the office publishes and distributes information booklets on IP topics with titles including: *Inventorship, Protecting Your Invention, Handling Your Invention, Marketing Your Invention* and *Should We Patent?*

Long-term benefits of having an intellectual property/technology transfer office
The long-term benefits of TTOs are enormous. Not only can the office assist in the protection of IPs generated within its institution, company

or organization, but also it can serve as a platform for generating revenues for research and development activities. In addition, a new company established through transfer of technology from the public sector to a private firm creates new job opportunities and enhances overall economic development

The success of any IP office depends largely on how well the technology transfer team interacts with each other. A clear line of communication between the parties involved is essential. For the office to remain successful and competitive, the team members must establish a positive reputation.

Information access

Appropriate information is key to IP use and management. Different parties require different types of IP information. Globally, as existing IP laws are modified to allow patenting of agricultural technologies such as genes and plant varieties, availability and access to current, worldwide patent information is becoming very important. A concerted international effort is needed to assist developing countries in accessing information on patents and other IP-related information.

Since ABSP began, the project has helped access information through various means (ABSP, 1997). Individuals from the collaborating countries have been able to access information through participation in ABSP-sponsored workshops (Maredia, 1995) and internship programs (Maredia *et al.*, 1996), and meetings of professional organizations such as the AUTM and BIO. Additionally, ABSP has helped establish internet services including e-mail. The project also publishes *BioLink*, a newsletter distributed to over 3000 individuals in more than 100 countries. Through these means, communication and networking within the biotechnology community has been greatly enhanced.

FUTURE DIRECTIONS IN CAPACITY BUILDING IN IPR

Capacity building in IP management is complex. These complexities require regional and global cooperation. Developing countries may learn from the experiences of developed countries. Capacity building success in any country will largely depend on how the parties involved in these complex issues communicate and work together. Addressing IPR issues and human resource development is critical. It is hoped that the developed nations with their wealth of IP management experience will assist developing countries so that true global interdependence can be achieved.

REFERENCES

ABSP (Agricultural Biotechnology for Sustainable Productivity) (1997) *Project Report to the Technical Advisory Group*. ABSP, Michigan State University, East Lansing, Michigan, USA.

Barton, J.H. and Bedford, B. (1993) Intellectual property rights: the ABSP intellectual property/patent workshop. *BioLink* Vol. 1 no. 4. ABSP Project, Michigan State University, East Lansing, Michigan, USA.

Bedford, B. and Maredia, K. (1994) *Proceedings of the AGERI & ABSP Workshop Series on Biosafety and Project Evaluation*, January 24–31, 1994, Cairo, Egypt, p. 31.

Ives, C. (1997) Implementation of plant variety protection in Morocco. *Seed World* 135, 34–36.

Maredia, K.M. (1995) ABSP workshop targets intellectual property rights in developing countries. *BioLink* Vol. 2 nos 2 and 3. ABSP Project, Michigan State University, East Lansing, Michigan, USA.

Maredia, K.M. and Dodds, J.H. (1994) Agricultural biotechnology for sustainable productivity: a unique initiative in USAID research and development. In: *Proceedings of the Third International Symposium on the Biosafety Results of Field Tests of Genetically Modified Plants and Microogramisms*, November 13–16, 1994, Monterey, California, pp. 449–452.

Maredia, K.M., Erbisch, F.E. and Dodds, J.H. (1996) Strengthening technology transfer framework in developing countries: a Michigan State University Internship Program. *Journal of the Association of the University Technology Managers* 8, 21–32.

APPENDIX 4.1. SELECTED INSTITUTIONS AND ORGANIZATIONS THAT OFFER IPR-RELATED INFORMATION AND TRAINING OPPORTUNITIES

US Universities
Professor John H. Barton, Stanford Law School, Crown Quad, Stanford, CA 94305-8610, USA. Tel.: +1-415-7232691; Fax: +1-415-7250253; E-mail: jbarton@leland.stanford.edu

Dr Frederic H. Erbisch, Office of Intellectual Property, Michigan State University, 238 Administration Bldg, East Lansing, MI 48824-1046, USA. Tel.: +1-517-3552186; Fax: +1-517-4321171; E-mail: erbisch@pilot.msu.edu

Dr William H. Lesser, Cornell University, Ithaca, NY 14853-1902, USA. Tel.: +1-607-2554595; Fax: +1-607-2559984; E-mail: WHL1@cornell.edu

Dr Harold C. Wegner, The George Washington University, Dean Dinwoodey Center, 2000 H Street NW, Washington, DC 20052, USA. Tel.: +1-202-994-4118; Fax: +1-202-9949446

Association of University Technology Managers (AUTM)
49 East Avenue, Norwalk, CT 06851-3919, USA. Tel.: +1-203-8459015; Fax: +1-203-8471304; E-mail: autm@ix.netcom.com; Web site: http://autm.rice.edu/autm

US Patent and Trademark Office (PTO)
Ms Mary Lee, USPTO, Washington, DC 20231, USA. Tel.: +1-703-3082359; Fax: +1-703-3052730; E-mail: mlee@uspto.gov

US Plant Variety Protection Office
Dr Marsha A. Stanton, Commissioner, Beltsville, MD 20705-2351, USA. Tel.: +1-301-5045518; Fax: +1-301-5045291; E-mail: Marsha_A_Stanton@usda.gov; Web site: http://www.usda.gov/ams/pvptitle.htm

World Intellectual Property Organization (WIPO)
34, chemin des Colombettes, Geneva, Switzerland. Tel.: +41-22-3389111; Fax: +41-22-7335428; Web site: http://www.wipo.org

International Union for the Protection of New Varieties of Plants (UPOV)
Mr Barry Greengrass, UPOV, Vice Secretary-General, 34, chemin des Colombettes, PO Box 18, 1211 Geneva 20, Switzerland. Tel.: +41-22-7309111; Fax: +41-22-7335428; E-mail: upov.mail@wipo.int

National Technology Transfer Center (NTTC)
316 Washington Ave, Wheeling, WV 26003, USA. Tel.: +1-304-2432455; Fax: +1-304-2432463; Web site: http://www.nttc.edu

Licensing Executives Society (LES)
1444 W. 10th Street #403, Cleveland, OH 44113, USA. Tel.: +1-216-2413940; Fax: +1-216-5669267.

Technology Transfer Society
23 N. Main Street, Franklin, IN 46131, USA. Tel.: +1-317-7383908; Fax: +1-317-7383980; E-mail: t2s@iquest.net; Web site: http//157.185.5.3/Default TTS.html

US Office of Technology Transfer
Dr Janelle Graeter, USDA/ARS, Beltsville, MD 20705, USA. Tel.: +1-301-5045676; Fax: +1-301-5045060; E-mail: jag@ars.usda.gov

International Service for National Agricultural Research (ISNAR)
Dr Joel Cohen, Project Manager, Intermediary Biotechnology Service, ISNAR, The Hague, The Netherlands. Tel.: +31-70-3496100; Fax: +31-70-3819677; E-mail: j.cohen@cgnet.com

Organization for Economic Cooperation and Development (OECD)
Ms Carliene Brenner, OECD Development Centre, 94 rue Chardon-Lagache, 75016 Paris, France. Tel.: +33-1-45249636; Fax: +33-1-45247943; E-mail: Carliene.Brenner@oecd.org

United Nations Industrial Development Organization (UNIDO)
Dr George Tzotzos, ICGEB/UNDIO, PO Box 300, 1400, Vienna, Austria. Tel.: +43-1-211-316180; Fax: +43-1-211-314336; E-mail: george@binas.unido.org

Country and Regional Case Studies

Egypt

Atef El-Azab

Academy of Scientific Research and Technology, 18 Midan El Misaha, Dokki, Giza, Egypt 12311

INTRODUCTION

This chapter covers Egyptian intellectual property rights (IPR) and their relation to the General Agreement on Tariffs and Trade (GATT). The chapter also discusses technology transfer in developing countries and its relation to Egypt's main economic resource – agriculture.

NATIONAL PERSPECTIVE

Current status of intellectual property laws

Inspired by the Paris convention for the protection of industrial property, which was passed on March 20, 1983, the Egyptian government issued a Law on Patents, Designs, and Industrial Models (No. 132, 1949) and its modification by Law no. 47 of 1981. The main features of this Law are the protection of new industrially exploitable inventions, new methods or processes of manufacture, and new applications of methods or processes already known.

The current Law states that no patent shall be granted for inventions relating to substances prepared or produced by chemical processes and intended for food or medicine. One exception is when the substances are prepared or produced by special chemical processes or operations. In the case of operations, the patent shall only cover such methods or processes of manufacture and not the substance itself.

The Law also states that a Register shall be held by the Academy of

Scientific Research and Technology to record all inventions and particulars relating thereto. The rights in an invention made by a worker or employee during working hours shall be vested in the employer. The term for a patent shall be 15 years from the date on which the application was made.

Some notable features of the Law are discussed below:

- Section 2 of this Law governs patent applications procedures. Should an application fulfil the conditions set forth in this section, the Patents Office at the Academy shall publish the invention as prescribed by the Executive Regulations. Any concerned person may object to the issue of the patent. The disposition is settled by a special judicial committee. One can contest the committee's decision before the Administration Court of the State Council. The patent is then issued by the appropriate minister.
- Section 3 deals with the assignment of the patent, its pledge and seizure.
- Section 4 deals with compulsory licenses and expropriation of patents for public utility.
- Section 5 deals with the termination of a patent and its revocation.

The second chapter of the Law is devoted to designs and industrial models. An explanatory memorandum affirms that food products are excluded from the Law's domain. This is because food products are not classed as inventions since they may pose risks to public health.

As from 1971, Presidential Resolution no. 2617 of 1971 vests the responsibility of patents in the Academy of Scientific Research and Technology, leaving, as before, the designs and industrial models to the Ministry of Supply and Internal Trade. Additionally, Law no. 14 of 1968, which was amended by Law no. 34 of 1975 and Law no. 38 of 1992, expanded protection to copyright and related aspects according to the Berne Convention of 1971.

Proposed changes in intellectual property rights laws

During the last decade, Law no. 132 of 1949 on Patents, Designs and Industrial Models was subject to many proposed changes. While the first changes had been completed before Egypt joined GATT, the last changes were completed only after GATT participation. That is, before GATT, the law was changed to encourage Egyptians to invent, to establish a scientific cadre and to introduce basic changes inspired from daily work and the public interest. The proposed changes were also intended to give employees, unless otherwise agreed upon, ownership rights in their inventions.

To make the law easier to implement, the utility model was also inserted (German Utility Model Law, 1968). This allowed a new tool's technical specifications to be included in the patent. The proposed changes created a link between the Patent Office and the factories, introduced a full inspection of the patent and widened the competence of the judicial committee to have full authority over all the conflicts arising out of the application of the law.

Another major change introduced food and pharmaceutical products as patentable subject matter and gave them full-term protection, as agreed under GATT. This reversed a trend and allowed a claim devoted to a non-naturally occurring composition, as in the US Supreme Court case: Commissioner of Patents and Trademarks v. Chakrabarty (case no. 79-136, June 16, 1980) (Barton, 1991).

After Egypt had joined GATT and the World Trade Organization (WTO), and after the conclusion of the Uruguay Round's Trade-related Aspects of Intellectual Property Rights (TRIPs), many amendments were added to the draft. These amendments included:

1. Increasing the patent's term to the new international norm – 20 years from the application's filing date.
2. Removing provisions that allowed the government to expropriate patents for public utility and instead permitting only compulsory licenses as outlined under TRIPs.
3. Asserting that the patent protection covers all fields of technology as outlined in Article 27 of TRIPs; the draft reads as follows:

> An invention patent shall be granted in accordance with the provisions of this Law, for every new innovative step feasible for industrial exploitation whether in connection with new industrial products, new industrial ways or means or new applications of industrial known ways or methods.

Here it is understood that the word 'industrial' means agricultural foodstuffs, medical drugs, pharmaceutical compounds, plant species, microbiological processes and their products.
4. Institution of the 'mail box' (Article 70/8), in which applications for patents not yet protected concerning pharmaceuticals and agricultural chemical products can be filed pending protection of these applications.
5. The grant of exclusive marketing rights (Article 70/9).
6. Protection of existing subject matter (Article 70/7).
7. The use without authorization of the right holder (Article 31).
8. Exhaustion (Article 6), which considers the owner of a patent to have exhausted his marketing rights in case he has already marketed his product in any other country.

The Egyptian Constitution (Article 151) also states that any international convention, in which Egypt participates and ratifies, is law. Presidential Resolution no. 72 of 1995, which made Egypt a WTO participant, was ratified by the House of Commons on 16 April 1995. The subsequent presidential ratification on 19 April 1995 made this Egyptian law.

A great debate ensued when deciding whether or not to delay for ten years the implementation of the pharmaceuticals and food products provisions of TRIPs. The industrial pharmaceutical sectors argued for the delay, stating that it would allow them time to face worldwide competition (Kabir, 1996; Reichman, 1993). Other sectors, such as agriculture, trade and culture, argued for immediate implementation because they wanted to encourage investment (Federation of Egyption Industries Report, 1996a,b). The final decision shall be submitted to the House of Commons by the Cabinet of Ministers. No decision has been made, the debate is still on.

Relationship between intellectual property rights and agriculture

Despite Law no. 132 of 1949 and its explanatory memorandum's statements that the word 'industrial' includes the use of patents in agriculture, the memorandum excludes inventions of foodstuffs and pharmaceutical compounds since the law allows only ten years' protection. Such an attitude, which covers genetic engineering, does not help promotion, development and investment, which Egypt greatly needs. Genetic engineering offers major tools for enhancing agricultural productivity and, hence, socio-economic development. Biotechnology research offers new tools and approaches to agricultural sustainability whereby food and fiber requirements may be met and the environmental quality enhanced. Egypt's failure to develop appropriate biotechnology applications and the inability to acquire technologies could deny her timely access to new advances (M.A. Madkour, Cairo, 1996, personal communication).

The new draft law overcomes these provisions because it expressly states that it applies to agriculture, foodstuffs, plant species and microbiological processes and their products. Therefore agriculture and its products are subject to protection as long as it is patentable subject matter.

Plant variety protection law

In Egypt there are no plant variety protection laws. According to TRIPs and the new draft law, the protection applies only when the invention meets the elements needed for a patent. Some efforts have been made in Egypt to adhere to the International Conventions for the Protection of New Varieties of Plants, held in Geneva on 23 October 1978 and its amendment on 19 March 1991.

TECHNOLOGY TRANSFER, COMMERCIALISM AND NATIONAL LINKAGES TO THIRD PARTIES

Licensing and other methods of technology transfer

Law no. 132 of 1949 is intended to organize the protection of patents and its procedures. The sole mention of licensing is made, as stated in the Paris Convention, only when the forfeit of patents and compulsory licenses occurs. This prevents abuse which results from the exclusive rights conferred by a patent.

Decades ago, Egypt aimed at wide industrialization and adapted various plans for this target. Because the acquisition of foreign technology was important, Egypt began executing licensing agreements with the outside world. During the 1960s and up to 1974, we had no freedom to choose the technologies needed in, for example, scientific, technical and economic sectors. The main target was building factories, supplied with old technologies from eastern European countries. It was rare to find a separate technology agreement. Most of the agreements included only a project study and report, engineering studies, supply of machinery and equipment, technical assistance, start-up tests, training and some legal terms. After the 1974 open-door policy, many agreements with western countries were made. However, many difficulties arose. Particularly, western countries now had opportunities to impose unfair terms and conditions; some of the terms and conditions included:

1. Demand for exorbitant licensing to be paid in a lump-sum and/or running royalties for long periods of time.
2. Obligating the licensee to buy the machinery, equipment, spare parts and raw material from the licensor at high prices.
3. Limiting the right of license to a specific project.
4. Rejecting the free exchange of any amendments between the two parties.
5. Limiting the right of the licensee to export the products as to certain countries and/or specific prices.
6. Imposing guarantee clauses not sufficient for the licensee.

During the 1970s, the UN General Assembly gave greater attention to technology transfer, and aimed at facilitating technology transfer to developing countries. The UN Conference on Trade and Development (UNCTAD) conducted many sessions between developed and developing countries. In an effort to issue an international code of conduct, including the fair terms of conditions for the transfer of technology to developing countries, Egypt participated in these sessions. However, no agreement was reached. The Academy of Scientific Research and Technology appointed a special committee to prepare a draft law for organizing the technology transfer contract. This draft law, which was inspired by the UNCTAD sessions and the Mexican law on licensing, has not been adopted. It is opposed by some sectors because they claim it restricts the open-door policy (El-Azab, 1995).

Egypt's General Organization for Industrialization is responsible for planning and encouraging industry projects, giving services and devices, and working as a consultant. They also evaluate and prepare technology agreements. This helps Egyptian firms sign agreements based on reasonable technical, financial, economical and legal conditions. These duties include: (i) informing the country about main technology sources; (ii) evaluating the draft contracts before signing; (iii) following up the technology supply phase; and (iv) helping to solve any problems that may arise between the two parties.

Other methods of technology transfer may arise by establishing collaborations with foreign firms' products. A new type of licensing in Egypt, called franchising, is also becoming available. A franchiser provides a standard package of products, systems and management services. Examples include McDonald's Restaurants, Kentucky Fried Chicken, Coca Cola and Pepsi Cola. In recent years, this kind of licensing has spread all over Egypt.

Commercialization investment

In order to enhance development in Egypt, a special law was issued to encourage investment. Law no. 43 of 1974, amended by Law no. 230 of 1989, specifies that the capital shall be deemed to mean seven items. These seven items include tangible assets such as patents and trademarks registered with member states of the International Convention for the Protection of Industrial Property or in accordance with the rules of international registration contained in the international conventions. Through investment, IPR play a major role in the development.

In reality, the adaptation of IPR in Egypt encouraged investors to put capital into various projects. This idea applies to all kinds of investment in fields such as industry, agriculture and pharmaceuticals, all of which Egypt needs greatly.

Although the percentage of Egyptian patents issued to Egyptians, compared with those issued to foreigners, is only about 5%, Egypt feels investment is urgent and needs to clear the way for investment by encouraging projects with more technology. In July 1996, the new government issued 24 new laws and regulations encouraging investment. This investment was encouraged by reducing or abolishing taxes and fees, and shortening or abolishing the procedures or formalities for the invested projects.

Links to international organizations

By Law no. 165 of 1950, Egypt participates in the following international agreements:

1. Paris Convention for the Protection of Industrial Property (March 1883 and its amendments).
2. Madrid Convention for the Registration of Trade and Industrial Marks (April 1891 and its amendments).
3. The Hague Convention for International Deposit of Trade and Industrial Marks (November 1925 and its amendment).
4. Madrid Convention for Geographical Indications (April 1891 and its amendments).

Egypt participated in the Paris Union and in the Bureau for the Protection of Intellectual Property (BIRPI). We also participate in the World Intellectual Property Organization (WIPO), which succeeded BIRPI. Using Article 4 of the agreement, WIPO gave assistance to Egypt by protecting its industrial property. Also, Egypt participates in the Food and Agriculture Organization (FAO) according to its rules.

Finally, another kind of cooperation exists between Egypt and the UN Industrial Development Organization (UNIDO). Egypt has been a member of UNIDO since its establishment. The purpose of UNIDO's Article 2 is to promote industrial development by encouraging the mobilization of national and international resources, and to assist in promoting and accelerating the industrialization of the developing countries. A particular emphasis is given to the manufacturing sector. The first Executive Director of UNIDO was an eminent Egyptian. At the outset of industrialization, Egypt received assistance from UNIDO through loans, technical assistance and training.

ILLUSTRATION OF IPR APPLICATION TO AGRICULTURE IN EGYPT

According to the patent records held at the Academy of Scientific Research and Technology, only one patent has been granted to Egyptian scientists in the field of agriculture, namely the patent granted to scientists from the Agricultural Genetic Engineering Research Institute (AGERI), Giza, Egypt. AGERI is a discipline oriented research institute within the Agricultural Research Center of the Ministry of Agriculture and Land Reclamation. This patent is on a biological insecticidal gene, isolated from a bacterium (*Bacillus thuringiensis*) indigenous to Egypt. It is the first of its kind to be obtained in Egypt for biotechnology and molecular biology related products (M.A. Madkour, Cairo, 1996, personal communication).

Concerning biodiversity laws, Egypt participated in the Biodiversity

Agreement concluded with UN Environment Programs in Rio de Janeiro, Brazil on 5 June 1992. With ratification on 6 May 1994, it became Egyptian law. No other laws have been issued.

REFERENCES

Barton, J.H. (1991) Patenting life. *Scientific American* 264, 40–46.

Carlos, M.C. (1993) *Prospects and New Dimensions of International Transfer of Technology, an issue paper.* UNIDO Experts' Meeting on Technology Transfer Trends, Vienna, Austria (issue paper).

El-Azab, A.M. (1995) Problems of Transfer of Technology through Licensing Agreements. *Journal of Council of State* 27, 1–15.

Federation of Egyptian Industries (1996a) *Report on Effects of Direct Application of TRIPs on the Pharmaceutical Industry in Egypt and its Comments.* Federation of Egyptian Industries Report, May 1996.

Federation of Egyptian Industries (1996b) *Patents, Economic Benefits, Myths and Reality.* Federation of Egyptian Industries, Seminar on Patency Rights, April 1996.

Kabir, O.A. (1996) *The TRIPs Agreement, Views of the Egyptian Public Business Sector Pharmaceutical Industry.* Seminar on Patency Rights, Cairo, Egypt (paper presentation).

Reichman, J.H. (1993) *Implications of the Draft TRIPs Agreement for Developing Countries as Competitors in an Integrated World Market.* UNCTAD discussion paper.

South Africa

6

Rosemary A. Wolson

Science and Technology Policy Research Centre, DPRU, Hiddingh Campus, University of Cape Town, Private Bag Rondebosch 7700, South Africa

INTRODUCTION

Concerted government policy of 'separate development' over many years, aimed at marginalizing the majority of the country's citizens, created a conspicuous dichotomy within South Africa, in which First World and Third World elements coexist. Many of the same conflicts experienced between nations of the North and nations of the South in the global arena, around issues such as free trade, technology transfer, intellectual property rights (IPR) and biodiversity, are therefore faced between different stakeholders within South Africa. In agriculture, this is particularly well illustrated, with the interests of the large-scale commercial concerns often at odds with those of small-scale traditional farmers.

Agriculture plays a vital role in the national economy, by contributing to food security and providing a significant source of export earnings. In addition, development of the informal farming sector offers opportunities for the upliftment of many South Africans who suffered oppression at the hands of the apartheid regime. The trend towards privatization of the technology which is necessary for future agricultural development dictates that an effective intellectual property (IP) regime be in place to ensure continued access to the best technology. In the South African context, the ideal regime must be flexible enough to meet the needs of both commercial and developing sector farming in order to maximize the benefits of agricultural development for all South Africans. It is with this in mind that South African IP law is therefore discussed.

© CAB INTERNATIONAL 1998. *Intellectual Property Rights in Agricultural Biotechnology* (eds F.H. Erbisch and K.M. Maredia)

The first section presents an overview of South African patent law and plant breeders' rights as the most important forms of IP protection available to innovators in the area of agricultural biotechnology. The extent to which the national IP regime affects technology transfer is then assessed. Finally, the impact of IPR is assessed on the commercial and traditional farming sectors respectively.

NATIONAL PERSPECTIVE

Current status of South African patent law

The information given here is based on Burrell (1986), Gerntholtz (1994), and personal communication from G. Tribe (patent attorney, Spoor and Fisher) and L. Parker (patent attorney, John and Kernick).

Background
In the region of 10,000 patents are filed annually in South Africa, approximately half of these being complete applications from other countries. The South African patent system, like that in Great Britain, offers the option of a two-stage application procedure. A provisional application may be filed up to 15 months before filing a complete application. This gives an effective date for the invention, from which priority can be claimed (i.e. novelty is only considered up to this date). A provisional patent specification requires less detail than a complete specification, and is cheaper. Alternatively, a complete application may be filed in the first instance. It is comparatively cheap for foreign concerns to file patents in South Africa, the entire cost of a patent application being R5000–8000 (equivalent to US$1100–1700 based on an exchange rate of R4.65 to the US$). This figure can be reduced if the specifications are already drawn up (L. Parker (patent attorney, John and Kernick) personal communication).

Patent legislation
South African IP legislation has historically been based on British law. More recently, developments in European law, especially in the European Patent Convention (EPC), have been followed. The respect with which IPR are treated in these jurisdictions has informed the South African law of patents. The patent legislation in place is well-established, providing the framework for a system which can function effectively.

Patents are governed by the Patents Act no. 57 of 1978, as amended (the Patents Act), and by Regulations made under the Act which deal with certain procedural matters. Patentability requirements in general are similar to those in British law and other jurisdictions. Section 25(1) of the Patents Act provides that a patent can be obtained for an invention which is novel, involves an inventive step (i.e. is 'unobvious') and which can be used or applied in trade or industry or agriculture. As in the corresponding British provision (Section 1(1) of the British Patents Act of 1977), an 'invention' is defined negatively, by stating what is not considered to be an invention for the purposes of the Act. A patent is granted for a term of 20 years from the date of filing the complete application, subject to the payment of prescribed renewal fees (Section 46(1) of the Patents Act).

The Patent Office
The South African Patent Office, which falls under the Department of Trade and Industry, is situated in Pretoria. Chief control of the Patent Office lies with the Registrar of Patents, appointed by the Minister of Trade and Industry. The Patent Office is relatively unsophisticated. It is a non-examining office equipped with neither the human resources nor the technology required for proper examination of the merits of a patent application. The Registrar therefore conducts only a formal examination, ensuring that all necessary procedural and administrative requirements have been satisfied. No novelty search is performed.

A consequence of having a non-examining Patent Office is that registration of a patent does not necessarily mean that the patent is valid. The patentee ought therefore to be more diligent in conducting novelty searches in order to ensure, as far as possible, that the invention covered by the patent is new.

The courts
The validity of a patent is ultimately determined by the courts, which have generally proved competent in this regard. A registered patent is *prima facie* valid. A party wishing to challenge the validity of a patent therefore bears the onus of proving invalidity in the courts. Patent litigation is instituted in the Court of the Commissioner of Patents. Commissioners are judges of the Transvaal Provincial Division (TPD) of the High Court of South Africa. Appeals to decisions of this court can be made to a Full Bench of the TPD, and thereafter to the Appellate Division.[1]

Because litigation takes place at the level of the High Court, it is very costly. As a result, small companies can rarely afford to litigate and even large companies with the necessary resources are often reluctant

---

[1] South Africa has a split bench, consisting of the Magistrates' Courts, which have limited jurisdiction, and the High Court, which consists of several provincial and local divisions, as well as the Appellate Division. The newly-established Constitutional Court has jurisdiction over constitutional matters. In the past, the right of appearance in the High Court was reserved for advocates. In 1995, however, the law was amended to give attorneys with certain qualifications the right of appearance in the High Court. Patent attorneys have always had the right of appearance in the Court of the Commissioner of Patents.

to do so, deterred by the high degree of technical expertise involved and the difficulty in finding expert witnesses.

Section 56 of the Patents Act provides for the granting of compulsory licences in cases where patent rights have been abused. Situations are listed which give rise to the presumption that an abuse of patent rights has occurred. An interested party may apply for a compulsory licence, which will be granted if the court determines that any of the listed situations exist. In practice, the tendency of the courts has been to adopt a strongly pro-patentee attitude in which patent rights are zealously protected and a resolute stand is taken against infringement and copying.

Proposed changes in IP law

It is accepted by government, the legal profession and industry that upgrading the Patent Office is not feasible in the foreseeable future, as the necessary resources are not available. The system generally operates effectively, bolstered as it is by strong legislation upstream and enforced by competent courts downstream. There are presently moves afoot to computerize certain sections of the patent office, which is expected to lead to improved record-keeping and search facilities. In addition, certain changes to the law have been recommended by innovators and legal practitioners to rectify the inequities that arise as a result of the fact that South African patent law is based on law developed for examining jurisdictions (G. Tribe (patent attorney, Spoor and Fisher), L. Parker (patent attorney, John and Kernick), C. Donninger (Bioclones), personal communications). These include:

- imposing an obligation on patentees to lodge with the Patent Office the results of all overseas examinations;
- limiting the time in which a patent can be amended (a period of four to five years after the patent has been granted has been suggested), in order to place the onus upon the patentee to amend a patent as soon as this becomes necessary;[2]
- introducing supplementary protection for pharmaceuticals and agrochemicals to give inventors more time in which to exploit their inventions and recoup research and development (R&D) costs, when regulatory procedures delay a product's entry into the marketplace.

---

[2] At present, if a patent application which has been accepted in South Africa is refused overseas, there is no obligation for the patentee to correct the South African patent. A party wishing to challenge the patent will therefore have to go to court to get it revoked or amended. The patentee can amend the patent to bring it within the bounds of validity even after litigation has begun. There are, however, certain consequences attached to failing to amend an invalid patent.

Further changes are envisaged when South Africa accedes to international treaties and acquires membership in international organizations.

The amendments necessary to harmonize South African IP law with the General Agreement on Tariffs and Trade (GATT) agreement on Trade-related Aspects of Intellectual Property Rights (TRIPs) are in the pipeline. An Intellectual Property Laws Amendment Bill, recently tabled in Parliament, has been drawn up to bring existing IP Acts into line with the TRIPs agreement. Most of the amendments are of a technical nature.

Accession to the Budapest Treaty on the International Recognition of the Deposit of Microorganisms for the Purposes of Patent Procedure is expected to take place during the course of 1997. Although the Patents Act provides for the deposit of samples of microorganisms, where a complete specification claims as an invention a microbiological process or product, and the microorganism concerned is not available to the public, the relevant section has yet to come into operation. Regulations for the deposit of microorganisms have been drafted (Regulation 28 *bis*), and will take effect in the near future, in order to comply with the requirements of the Treaty. Moves are under way to set up an International Depository Authority in South Africa (Patent Office, personal communication). In practice, deposit usually takes place for microbiological inventions, notwithstanding the fact that it is not yet mandatory, because in the absence of a deposit, it is difficult to satisfy the sufficiency requirement for patentability (G. Tribe (patent attorney, Spoor and Fisher), J. Thomson (Head of Department of Microbiology, University of Cape Town), personal communications).

Biotechnology patents in South Africa

While it is estimated that fewer than 5% of home-grown patents filed in South Africa are for biotechnological inventions, a significantly higher proportion of foreign applications are biotechnology-related (G. Tribe (patent attorney, Spoor and Fisher), personal communication). The Patent Office has no statistical breakdown of patent applications in different fields.

The patentability of biological inventions is dealt with in section 25(4)(a) of the Patents Act, which adopts similar wording to the corresponding EPC provision (EPC Article 53(b)). It states that, 'A patent shall not be granted for any variety of animal or plant or any essentially biological process for the production of animals or plants, not being a microbiological process or the product of such a process.' While microbiological processes and their products are therefore clearly patentable, exactly what constitutes a 'microbiological process' or a product thereof, and what the situation is regarding the patentability of other living material, is less clear, because the Act does not define key terms such as 'variety', 'essentially biological' or 'microbiological', and our courts have not been called upon to interpret these terms. Guidance is therefore obtained from other jurisdictions whose legislation contains similar provisions, the European Patent Office (EPO) Guidelines for Examination being a particularly useful source in this regard. (It must be borne in mind, though, that this area of law is not clear-cut, even in those jurisdictions whose lead South Africa tends to follow.) A 'microbiological process or the product of such a process' is therefore expected to include microorganisms as well as processes involving their use and utility.

Van der Merwe (1993) submits that biological inventions are intrinsically patentable subject matter and that their exclusion from patent protection is based on extrinsic factors and therefore limits existing rights, which suggests that this provision ought to be interpreted restrictively, so as to exclude only natural and uncontrolled processes. Nontrivial human technical intervention would therefore ensure that a process was not 'essentially biological', even if it were partially biological. Genetic modification of plants and animals would thus not be considered 'essentially biological' according to this interpretation (Van der Merwe, 1993; Tribe, 1996).

In summary, the Act therefore appears to offer protection for microbiological organisms and processes, as well as for processes producing transgenic plants or animals, and for the products of such processes (provided there is a sufficient degree of human intervention), unless the plant or animal product of such process is a variety.

It is interesting to note that a precedent for the patenting of plant material in South African law can be found in the repealed Patents Act no. 37 of 1952. Prior to the introduction of the Plant Breeders' Rights Act no. 22 of 1964, new plant varieties were patentable. It can be argued that this is evidence of the patentability of living material under South African law. An alternative interpretation, however, holds that the express provision for the patenting of plant material implies that inventions in the field of animal life are not patentable, such inventions having been excluded from the scheme of the repealed Patents Act (Van der Merwe, 1993). Nevertheless, this debate is likely to be less important than the persuasive authority of other jurisdictions, in the event of South African courts being called upon to decide the question of the patentability of living material, taking into account that: (i) the Act concerned has been repealed and (ii) the relevant South African legislation now in force is based on European legislation.

Patents have been granted by the South African Patent Office for various biotechnological inventions, including genetically modified microorganisms, plants and animals. Whilst none of these patents has been challenged in the courts, and their validity could therefore still be brought into question, it is generally accepted from a practical perspective that at least some of these patents are valid. The fact that biotechnology companies continue to file patents for inventions dealing with living material is evidence that there is a degree of confidence in the protection offered by the legislation (G. Tribe (patent attorney, Spoor and Fisher), personal communication).

Plant variety protection

Legislation (Dold, 1982; Van der Walt, 1996)
Plant variety protection is governed by the Plant Breeders' Rights Act no. 15 of 1976, as amended.[3] The Plant Breeders' Rights Amendment Act no. 15 of 1996 was enacted to bring the South African legislation into compliance with the 1991 revisions of the International Union for the Protection of New Varieties of Plants (UPOV). A plant variety is eligible for protection if it is novel, distinct, uniform and stable, and its denomination (generic name) complies with prescribed requirements. The novelty requirement is met where important new characteristics are brought about by alteration of existing characteristics through selection and breeding. The term of protection varies according to the type of plant for which protection is sought, ranging from 20 to 25 years (prior to the 1996 amendments, the minimum term of protection was 15 years). In terms of Regulations under the Act, only varieties of listed plant species could be registered (this list being changed from time to time). Protection will be extended, however, to all plant genera and species, in line with the 1991 UPOV requirements.

The owner of a plant breeder's right has the exclusive right to exploit a protected plant variety and can exclude others from producing, selling, importing or exporting its propagating material. Private, non-commercial or experimental use of a protected variety for further breeding does not fall within the ambit of the protection conferred by the Act.

Changes brought about by the amending Act include the introduction of the concept of an 'essentially derived variety', commercial use of which requires the consent of the owner of the initial protected variety, and the extension of protection to harvested material in cases where the breeder is unable to obtain remuneration rightfully on the propagating material.

Application (Van der Walt, 1996)
Plant breeders' rights are administered by the Directorate of Plant and Quality Control of the Department of Agriculture. Table 6.1 shows a

---

[3] The Plant Breeders' Rights Act is complemented by the Plant Improvement Act no. 53 of 1976, which introduces a system of certification for certain plants deemed to be of economic importance, to maintain quality control within the seed industry.

Table 6.1. Total valid plant breeders' rights in South Africa in 1994.

Crop group	Residents	Non-residents	Total
Agronomic/forage	188	74	262
Fruit	106	34	140
Ornamentals	54	218	272
Vegetables	68	62	130
Total	416	388	804

breakdown by crop group of the 804 plant breeders' rights which were in force in South Africa in 1994. More than half of these were held by South Africans, while 180 South African applications for new varieties were lodged in the European Union during that year.

The fact that plant breeders' rights have been limited to varieties of prescribed types of plants has probably led to protection under the Act being sought less frequently than it might have been, up to now. Nonetheless, plant variety protection has been beneficial to private enterprise, farmers and consumers in a number of ways:

1. It has been cited as an important factor in stimulating private investment in plant breeding, which has increased considerably since the introduction of plant breeders' rights (Van der Walt, 1996).

In 1993, private companies and cooperatives accounted for almost 60% of total plant breeding expenditure (Van der Walt, 1996). This figure is likely to increase with the decrease of funding levels for public institutions.

2. It gives local breeders the opportunity to benefit from wider access to new varieties released internationally.

3. Royalties on protected varieties are an important source of funding for public research institutions, where levels of government funding are declining.

4. Farmers and consumers benefit from increased crop yield and improved crops resulting from new varieties.

One of the biggest deficiencies in the system of plant breeders' rights is the inadequacy of enforcement mechanisms. Enforcement is by civil litigation between breeder and infringer, and is not considered to be very effective against seed piracy.

TECHNOLOGY TRANSFER AND INVESTMENT IN BIOTECHNOLOGY R&D

South Africa's IP regime, although not without flaws, is considered adequate in most respects. Patents have always enjoyed strong protection in South Africa, while the problems associated with the lack of protection conferred upon foreign trademarks in the past are being addressed as South Africa strives to make its markets attractive to the international community. In 1995, South Africa was placed on a 'watch list' by the Office of the United States Trade Representative. This was a result of certain local businesses having exploited the trademarks of foreign corporations who had failed to register their trademarks in South Africa, or to use or appoint a registered user for their marks, when these corporations were prevented from investing in South Africa due to international sanctions. The need to attract foreign investment has ensured that this situation no longer pertains, as is illustrated by a high-profile case involving McDonald's Corporation. Recent amendments to the law of trademarks provide, amongst other things, for the recognition of well-known trademarks, and the 1996 decision of the Appellate Division of the High Court of South Africa in that case has ensured that the threshold of recognition required for a mark to be considered 'well-known' is not too high. The outcome of this case, which saw the decision of the lower court reversed, has been welcomed as a signal that South Africa is committed to the protection of IP.

Overseas biotechnology companies appear to be satisfied with the protection offered by the South African patent system (i.e. the legislation in place and its enforcement), and are therefore not discouraged from engaging due to fears that adequate protection will not be received for patented technology. However, this alone is insufficient to ensure that biotechnology originating overseas is readily available to South African R&D concerns, because leading technology is often not for sale, or carries too high a price. While the need for an effective system of IP protection is not in question, in certain cases, the strong recognition given to patent rights can impede the development of the local industry. Furthermore, non-IP-related barriers exist which curtail South Africa's ability to attract foreign technology and investment in biotechnology R&D.

IP-related factors

Pro-patentee stance of the courts

The pro-patentee stance of South African courts with regard to the refusal to grant compulsory licences in the case of alleged abuse of patent rights is one factor which allows the South African IP regime to be used as a barrier to technology transfer. Patent practitioners and others involved in the biotechnology industry have expressed the view that if the courts were to take a more lenient approach in this regard, local industry could be stimulated. This need not unduly dilute IPR, provided the courts adhere consistently to suitable criteria when interpreting whether an abuse of patent rights has in fact occurred.

High royalties

The utilization of what have become standard laboratory techniques in the generation of a commercial product is of great concern to agricultural biotechnologists. Most of these techniques are patented in South Africa, but permission has been granted for many of them to be used free of charge for research purposes. However, once these techniques are incorporated into the development of a product, licensing agreements will in all probability have to be entered into with the patentees. It is commonly believed that the terms of such agreements will include the payment of exceptionally high royalties (estimated to be between a third and a half of all profits) which will inflate the price of the ensuing product to the extent that it is no longer cost-effective, notwithstanding the advantages it confers (Johan Brink (ARC) and Johan Burger (CSIR), personal communications). This problem has not yet arisen in practice, as product development is not sufficiently advanced, but is likely to occur in the near future.

Membership of international conventions

Technology transfer might be inhibited to some degree by the fact that South Africa is not a signatory to certain international conventions, but this is a temporary obstacle in view of the fact that accession to the major conventions will be taking place in the near future.

Other factors that affect biotechnology transfer

Exchange control regulations

Exchange control regulations limit royalties that can be paid out of the country to between 4% and 7%. Foreign companies wanting higher royalties are therefore deterred from entering into license agreements with South African concerns, as the balance of such royalties must remain in South Africa.

The administrative and procedural obstacles that arise as a result of the exchange control regulations are a further impediment to technology transfer. Even where no money leaves the country at the time of entering into an agreement, Reserve Bank approval must be obtained, often on the basis of speculative data, the collation of which is time-consuming and leads to potentially costly delays.

The government has repeatedly stated its commitment to the scrapping of exchange control regulations as part of its efforts to create an economic climate conducive to investment. The failure to impose a time-frame for this, however, is a matter of concern for potential foreign investors.

Global trade climate

In the prevailing global trade climate, where it is becoming increasingly difficult for outsiders who do not have the best technology to enter and penetrate world markets, South Africa faces a major stumbling block as a result of having been isolated from the world community for so long. For many years, R&D concerns were denied the opportunity to enter into joint ventures, which are an important means of gaining access to new technology and to new markets.

Uncertainty regarding South Africa's stability

Foreign companies are concerned about the country's economic and political stability and foreign investment is dominated by a 'wait-and-see' attitude. Until South Africa is able to instil greater confidence in investors, investment will continue to be on a limited scale.

Lack of biosafety legislation

The lack of biosafety legislation has possibly affected biotechnology transfer to some degree, with overseas companies reluctant to conduct trials or invest in other projects that involve genetically modified organisms (GMOs) in South Africa. The Department of Agriculture has addressed the need for such legislation with a Draft Genetically Modified Organisms Bill, to regulate all stages of the handling of GMOs, including importation, production, release and distribution. It is anticipated that the Bill will be passed into law in the forthcoming parliamentary session.

In the absence of biosafety legislation, regulation of the handling of GMOs has been taking place under the auspices of the South African Committee for Genetic Experimentation (SAGENE), an advisory body of experts in different fields relevant to GMO work. SAGENE has drawn up a code of conduct in terms of which an individual or organization importing, releasing or carrying out research on GMOs undertakes to consult with SAGENE and abide by SAGENE's recommendations. Despite the fact that there is no legal obligation to conduct GMO releases through SAGENE, there is no evidence of any releases having taken place in the country without SAGENE's involvement, and SAGENE has overseen a number of successful field trials (Morris, 1995).

International linkages

South Africa is a signatory to the Paris Convention for the Protection of Industrial Property, it is a member of the World Trade Organization (WTO) and it has been a member of UPOV since 1977. Its status as the only African country to have joined UPOV, and the fact that there was so little opposition on the part of the developing agricultural sector when the 1991 revisions were ratified, are an indication of the influence wielded by the formal seed sector (Fakir, 1996).

Moves are under way for South Africa to join the Patent Cooperation Treaty (PCT), but it is difficult to predict a time-frame for this, an important constraint in this regard being the training of personnel to administer the Treaty provisions (Patent Office, personal communication).

South Africa is likely to join the African Regional Industrial Property Organization (ARIPO), although a final decision in this regard has yet to be made. Relatively few applications are filed at ARIPO due to a lack of confidence in the organization, whose infrastructure is perceived to be unsophisticated. Furthermore, the costs of a regional patent office are considerable. Most patenting activity in the region takes place in South Africa; other member states would arguably stand to benefit from the arrangement more than South Africa would. Notwithstanding these drawbacks, South Africa is entering the markets in a number of member states, and will therefore require IP protection there (L. Parker (patent attorney, John and Kernick) and J. Morris (AECI Ltd), personal communications). In November 1995, South Africa ratified the 1992 Convention on Biological Diversity.

Because of South Africa's isolation during the apartheid years, linkages with international organizations were few and far between, and such relationships were often 'unofficial'. Now that South Africa has rejoined the international community, several initiatives have been and continue to be made to remedy this situation, and South Africa's international contacts in agricultural biotechnology, as in other areas, are increasing. The following list is not exhaustive:

1. South Africa has collaborated over the years with the International Seed Testing Association (ISTA), the International Seed Trade Federation (FIS), the Organization for Economic Cooperation and Development (OECD), UPOV and the International Plant Protection Convention (IPPC), for the purposes of furthering plant and seed production and quality control.
2. More recently, with ratification of the Undertaking on Plant Genetic Resources, collaboration has taken place with the International Plant Genetic Resources Institute (IPGRI).
3. South Africa has been a member of the United Nations Food and Agriculture Organization (FAO) since 1994.
4. Regional cooperation in the conservation and development of agricultural resources has taken place with certain Southern African countries, through the Southern African Regional Commission for the Conservation and Utilization of the Soil (SARCCUS) (Department of Agriculture, 1996).

APPLICATION OF IPR TO AGRICULTURE

Farming practices in South Africa are diverse, ranging from the large-scale commercial sector to small-scale operations and subsistence farming (Mogford, 1996).

Commercial sector agriculture

Agricultural and forestry products (in processed and unprocessed form), including wool, maize (corn), sugar, citrus, deciduous fruit, wine and paper, usually make up approximately one third of the country's total export earnings (Department of Agriculture, 1996). Evidence of the success of the commercial farming sector can be found in the fact that South Africa is largely self-sufficient in a number of crops, including maize (by far the dominant crop), wheat, oilseed and sugar, despite a number of constraints. For example:

- only 14% of the land is estimated to be suitable for crop cultivation (and as little as 3% is considered to be high potential land);
- crop yield is dependent upon rainfall, which is highly variable from year to year;
- all of the major food crops are exotic species which have had to be adapted for local conditions and whose sustainability is dependent upon continued access to imported germplasm (Webster and Koch, 1995; Department of Agriculture, 1996; Koch and Webster, 1996; Moss, 1996).

Crop improvement programs are therefore driven by the need to overcome these limitations, and plant biotechnology initiatives are focused largely on drought tolerance, viral, fungal and insect resistance and herbicide tolerance. Good infrastructure and a solid, albeit small, base of expertise exist. The Agricultural Research Council (ARC), and the Council for Scientific and Industrial Research (CSIR), both statutory science councils with partial government funding, are the most active research organizations engaged in plant biotechnology work. A number of university research groups also have plant genetic engineering projects under way. At the commercial level, in-house plant biotechnology R&D facilities have been set up by both the chemical and sugar industries (Webster and Koch, 1995). Some of the current research projects in plant genetic engineering are listed in Table 6.2.

All of the main South African companies engaged in crop breeding have relationships with major multinationals. With respect to most crops, the South African market is comparatively limited; such relationships are therefore vital as a source of funding, expertise and international germplasm for the main food crops (Mogford, 1996). In Table

Table 6.2. Current research into plant genetic engineering applications for crop improvement.

Trait	Crop	Research group
Drought resistance	tobacco, cotton, potato, maize	ARC-VOPI
Insect resistance	cotton	ARC-VOPI
	maize, sorghum	CSIR
	wheat	UOFS
	sugarcane	SASEX
Herbicide resistance	melon, tobacco	ARC-VOPI
	maize, soybean, sunflower	CSIR
Viral resistance	potato, watermelon	
	Ornithogalum, tomato	ARC-VOPI
	tobacco	ARC-Infruitec, UCT
	maize	UCT
Fungal resistance	tomato, strawberry	ARC-Infruitec/VOPI
	wheat, maize, sorghum	CSIR
	cotton	RAU
	pine, stone fruits	US
Shelf-life	cut flowers	UN

ARC-VOPI, Agricultural Research Council – Vegetable and Ornamental Plant Institute; ARC-Infruitec, Agricultural Research Council – Institute for Fruit Technology; CSIR, Council for Scientific and Industrial Research; RAU, Rand Afrikaans University; SASEX, South African Sugar Association Experimental Station; UCT, University of Cape Town; UN, University of Natal; UOFS, University of the Orange Free State; US, University of Stellenbosch.

6.3, partnerships which exist between local and international companies for different products are shown.

The South African National Seed Organization (SANSOR), a non-governmental organization representing 92 companies, is responsible for coordinating research, production and marketing for the major seed companies. Other functions carried out by SANSOR include seed certification, variety testing and licensing of new varieties.

Developing sector agriculture

Information regarding the informal farming sector is limited. Traditional farming is less prevalent than in most other African countries, largely on account of apartheid policies which resulted in the forced removal of numerous communities from the land they occupied (Dakora, 1996; Department of Environmental Affairs and Tourism, 1996). Farming in developing communities is a low-technology enterprise, usually taking place on small areas of private or communal land. Cattle and goat herds

Table 6.3. Selected South African companies involved in plant biotechnology projects, showing international partnerships.

South African company	Overseas partner	Product
AgrEvo	AgrEvo	herbicide-resistant maize and soybeans
Carnia	Asgrow	soybeans
Carnia	Cargill	canola and maize
Clark Cotton	Delta Pine & Land Co.	*Bt* transgenic cotton
Hadeco	–	micropropagation of disease-free ornamental bulbs
Mayfords	Yes	genetically improved tomatoes
National Chemical Products	Yes	biocontrol and fertilizer
Pannar	Yes	modification in sunflowers
PHI Genetics	Pioneer Hi-Bred	maize, soybeans, sunflower and sorghum
Sensako	DeKalb Genetics Corp.	virus-resistant maize, pest-resistant wheat and molecular markers in breeding program

often compete with agronomy. It seems that the situation where crops are sold to generate cash is more common than subsistence farming (Webster and Koch, 1995). Commercial crops, attractive to traditional farmers because they are easy to grow and give a greater yield, have been displacing traditional crops over a prolonged period of time, resulting in erosion of the traditional knowledge base and an ongoing loss of genetic diversity (Dakora, 1996; Fakir, 1996). The ARC is one institution investigating traditional and neglected crops (J. Brink (ARC-VOPI), personal communication), but similar initiatives are few and far between.

In the past, efforts of the formal plant breeding sector, targeted almost exclusively at commercial farmers, have neglected the traditional farmers, who consequently lack access to seed distribution networks. At the same time, local seed systems characteristic of most African countries, through which seed is selected, maintained, replaced and exchanged within communities of small farmers, are also lacking in South Africa.

This is probably one reason why traditional farmers in South Africa are responsible for less genetically diverse germplasm than traditional farmers elsewhere, but there are also more fundamental reasons for this. One factor suggested in this regard is that, because South Africa has been populated for a relatively shorter time than most other parts of the southern hemisphere, indigenous peoples have had less time to breed crops and select crop genetic material. An additional reason lies in the fact that South Africa is neither a center of origin nor of diversity for any major crop species. Preliminary studies seem to indicate, however, that traditional farmers have contributed to the promotion of genetic diversity in crops (Rachel Wynberg, personal communication). Further research in this area, which has been neglected in the past, is clearly necessary to learn more about the nature of this contribution, and to assess how compensation might be granted, for example by introducing a system of farmers' rights (Moss, 1996; Department of Environmental Affairs and Tourism, 1996). Some form of legal recognition is also called for in respect of traditional knowledge regarding non-agricultural aspects of national biodiversity. South Africa, ranked as the world's third most biologically diverse country according to an index derived by the World Conservation Monitoring Centre, contains an immense wealth of indigenous genetic diversity and a very high proportion of endemic species, especially of vascular plants. Their potential lies especially in medicinal use and horticultural applications, and as forage plants.

While government acknowledges the value of traditional knowledge and the need to reward its holders (e.g. Department of Environmental Affairs and Tourism, 1996), the implementation of concrete mechanisms to facilitate this is a long way off. The existing IP law was not designed to embrace the concept of collective property rights underlying traditional knowledge; the development of a *sui generis* system therefore seems to be necessary. South Africa has many strengths, including its wealth of genetic resources, a developed agricultural sector, infrastructure and expertise in biotechnology and a functional regime in place for conventional IP protection, which can be built upon to widen the scope of the existing IP system to take into account both the requirements of the world trade order and the need for recognizing the innovative contribution of the informal farming sector (Moss, 1996). Positive, practicable suggestions as to how this should be done, however, are in short supply. Industry and the legal profession caution against measures which might result in the dilution of conventional IPR, to the detriment of technological and industrial progress.

CONCLUSION

The promotion of technological and industrial progress is, after all, one of the main goals of any system of IPR. The fact that the commercial sector is well-served, for the most part, by the existing regime would seem to indicate that it is fulfilling its intended function. However, in the context of South Africa, any system that continues to discriminate against those who suffered institutionalized discrimination in the past cannot be condoned. By failing to recognize traditional knowledge as IP, the conventional IP system is to some extent discriminatory. It is accepted, however, that such a system is a prerequisite in the prevailing global trade climate, and that tampering with it would diminish the country's competitiveness in international markets. Parallel legislative and policy measures are therefore required to ensure that developing communities are not sidelined by technological progress, and that traditional knowledge is recognized as a valuable resource and is conserved and exploited, especially for the purposes of developing the informal rural sector.

REFERENCES

Burrell, T.D. (1986) *South African Patent Law and Practice*, 2nd edn. Butterworths, Durban.

Dakora, F.D. (1996) Use of plant genetic resources by traditional farmers. In: *Proceedings of Land and Agriculture Policy Centre Genetic Conservation Workshop, Johannesburg, 19–20 March 1996*.

Department of Agriculture (1996) *Country Report for South Africa*, compiled for the UN FAO International Conference and Programme for Plant Genetic Resources (ICPPGR). Government Printer, Pretoria.

Department of Environmental Affairs and Tourism (1996) *Green Paper on the Conservation and Sustainable Use of South Africa's Biological Diversity*. Government Printer, Pretoria.

Dold, D.M. (1982) Plant breeders' rights. *De Rebus* 475, 178.

Fakir, S. (1996) Local knowledge and intellectual property rights in South Africa: between collective sharing and privatisation of knowledge. Land and Agriculture Policy Centre (in press).

Gerntholtz, R. (1994) *Basic Guide to the Law of Patents*, 3rd edn. Allegretto (Pty) Ltd, Cape Town.

Koch, M. and Webster, J. (1996) *Biotechnology in South Africa*. Foundation for Research Development (in press).

Mogford, D. (1996) The use of genetic resources in agricultural crop production and the seed industry. In: *Proceedings of Land and Agriculture Policy Centre Genetic Conservation Workshop, Johannesburg, 19–20 March 1996*.

Morris, E.J. (1995) A report on activities carried out by SAGENE in 1995. *Biosafety Network Newsletter* 1(2), 5.

Moss, M. (1996) *The Application of Intellectual Property Protection to Biodiversity and Agriculture in South Africa: Consequences, Concerns and Opportunities*. Working Paper 36, Land and Agriculture Policy Centre.

Tribe, G. (1996) Patent rights for biological organisms in South Africa. In: *Proceedings of Land and Agriculture Policy Centre Genetic Conservation Workshop, Johannesburg, 19–20 March 1996*.

Van der Merwe, A. (1993) Die beskerming van vooruitgang op die gebied van lewende materie in Suid-Afrika. *Tydskrif vir die Suid-Afrikaanse Reg* 1, 139–150.

Van der Walt, W.J. (1996) Plant breeders' rights in South Africa. In: *Proceedings of Land and Agriculture Policy Centre Genetic Conservation Workshop, Johannesburg, 19–20 March 1996.*

Webster, J. and Koch, M. (1995) South African plant biotechnology. In: Altman, D.W. and Watanabe, K.N. (eds) *Plant Biotechnology Transfer to Developing Countries.* RG Landes Company, Austin, Texas, pp. 75–83.

7

Australia

Michael Blakeney

Asia Pacific Intellectual Property Law Institute, School of Law, Murdoch University, Murdoch, WA 6150, Australia

CURRENT STATUS OF INTELLECTUAL PROPERTY LAWS

Australia is a federation, or Commonwealth, of eight States and Territories, each with its own court system and parliament. Overarching the Commonwealth is a national Parliament which has paramount legislative power on subjects which are listed in the federal Constitution. The Constitution dates from the turn of the century. Section 51 (xviii) of the Constitution confers power upon federal Parliament to legislate with respect to 'copyrights, patents of inventions and designs and trade marks'. A potential problem for the federal Parliament is that new categories of intellectual property right, such as plant breeders' rights, are not listed in section 51 (xviii) of the Constitution. At the apex of the Australian court system is the High Court of Australia. Recent High Courts have dealt with this problem by permitting federal legislation pursuant to the 'external affairs' power in section 51(xxix) of the Constitution. Thus upon Australia's accession to the International Convention for the Protection of New Varieties of Plants (the UPOV Convention) in 1989, federal legislation on this subject could be constitutionally grounded.

Relevant federal intellectual property (IP) legislation includes the Designs Act 1906, Copyright Act 1968, Circuit Layouts Act 1989, Patents Act 1990, Plant Breeder's Rights Act 1994, and Trade Marks Act 1995. Australia's IP statutes were originally largely re-enactments of equivalent British statutes, but over the years these have been refashioned in line with Australia's national requirements, as well as its international IP obligations. Another significant federal statute with implications for IP is the Trade Practices Act 1974, which contains a comprehensive code proscribing unfair competition. Thus infringement cases typically combine claims under the relevant intellectual property statute and allegations of 'misleading or deceptive conduct' in breach of section 52 of the Trade Practices Act 1974.

Since its creation in 1976, the Federal Court of Australia exercises jurisdiction in relation to federal statutes. Before this date, jurisdiction in IP matters was exercised by State and Territory Courts. Since 1976, litigants have the choice of initiating litigation in State or Federal Courts, although, in practice, the Federal Court of Australia is increasingly becoming the preferred forum. Confidential information and trade secrets are protected under common law, through actions in the State and Territory Courts.

Australia is a signatory to all the major IP conventions[1] and is a member of the World Intellectual Property Organization and the World Trade Organization (WTO).

The industrial property statutes are administered by the Australian Industrial Property Organization, with the exception of the Plant Breeder's Rights Act 1994, which is administered by the federal Department of Agriculture. Copyright matters fall within the jurisdiction of the federal Attorney-General's Department.

RECENT AND PROPOSED CHANGES IN INTELLECTUAL PROPERTY RIGHTS LAWS

The Commonwealth Parliament has been progressively reviewing all of Australia's IP laws. The federal Bureau of Industry Economics has been commissioned to report on the legislative protection of each category of industrial property right. The Patents and Trademarks Laws were comprehensively updated by new laws in 1990 and 1995, respectively. The Plant Varieties Act 1987 was replaced by the Plant Breeder's Rights Act 1994. In October 1994 the Australian Law Reform Commission released a discussion paper setting out proposals for a comprehensive revision of the current industrial designs law and this has been endorsed and amplified by a subsequent report issued by the Bureau of Industry Economics (1995). Also in October 1994 a Copyright Convergence Group, appointed by the Minister for Justice, released a report proposing a new copyright regime for the commercial transmission of copyright materials by electronic means (Copyright Convergence Group,

[1] In relation to patents and plant varieties Australia has acceded to the Paris Convention for the Protection of Industrial Property 1883, Patent Cooperation Treaty 1970, Budapest Treaty on the International Recognition of the Deposit of Microorganisms for the Purposes of Patent Procedure 1977, Strasbourg Agreement Concerning the International Patent Classification 1971 and the UPOV Convention 1961.

1994). This report has been referred to the Copyright Law Review Committee, which has been asked to examine the ways in which it might be implemented as part of the Committee's brief to simplify the legislation.

Australia was a founding member of the WTO and, consequently, a signatory to the Agreement on Trade-related Aspects of Intellectual Property Rights (TRIPs Agreement) (Blakeney, 1996a). In conformity with its obligations under the WTO Agreement, the Commonwealth Parliament has passed three Acts to bring the country's IP laws into conformity with the TRIPs Agreement (Blakeney, 1996b). These were the Patents (WTO Amendments) Act 1974, Copyright (WTO Amendments) Act 1994 and the Trade Marks Act 1995. Of these statutes, the Patents and Trade Marks (WTO Amendments) Acts will probably have the greatest significance for agriculture. The principal changes effected by these Acts and the other recent IP laws are detailed below.

Patents Act 1990

On 30 April 1991 a new law, the Patents Act 1990, repealed and replaced the previous Patents Act 1952. The new law was passed as a result of a review of the patents legislation by the Industrial Property Advisory Committee (1984). The new Act was designed to simplify procedures to make patenting more accessible to non-experts. Under the new law the assessment of novelty and inventiveness was changed from the benchmark of national prior art to a standard of global prior art and publication. Additionally, a 'whole of contents' approach is taken to the assessment of novelty, in that the entirety of a specification will be examined and not just the claims made in an earlier application.

A major change effected by the new legislation is in relation to the right of exploitation of patents. Under section 69 of the 1952 Act a patentee had the exclusive right to 'make, use, exercise and vend' the invention. Section 13(1) of the 1990 Act defines the right given by a patent as 'the exclusive rights, during the term of the patent to exploit the invention and to authorize another person to exploit the invention'. The term 'exploit' is defined in Schedule 1 to the Act to include: '(a) where the invention is a product – make, hire, sell or otherwise dispose of the product, offer to make, sell, hire or otherwise dispose of it, use or import it, or keep it for the purpose of doing any of those things; or (b) where the invention is a method or process – use the method or process or do any act mentioned in paragraph (a) in respect of a product resulting from such use.'

Finally, the 1990 Act adds a new form of infringement, consisting of the supply of a product, where use of the product would be an infringement, provided that use is the only reasonable use of the product and that use is in accordance with any instructions, inducement or advertisement given or published by the supplier, or, in the case of a non-staple product, that use is the one to which the supplier had reason to believe the receiver would put it (Patents Act 1990, section 117).

Patents (WTO Amendments) Act 1994

Patent term

The term of a standard patent granted after 1 July 1995 is extended by the amending Act to 20 years, instead of the previous patent term of 16 years, with the possibility of a four year extension for pharmaceutical patents. Additionally, patents granted under the previous law which were due to expire after 1 July 1995 are to be extended to a 20 year term.

Compulsory patent licences

In the situation where a person, prior to the amending legislation, made a significant investment in anticipation of the expiry of a patent after 16 years, the amending Act provides for the grant of a compulsory license to that investor. The preconditions for this license are that: (i) the investment was made in good faith prior to 1 October, 1994; (ii) no action done by the applicant in preparation for the exploitation infringed the patent; and (iii) the applicant tried for a reasonable period, but without success, to obtain a license from the patentee on reasonable terms. A similar extension is effected for licenses which are due to expire at the end of the 16th year of the term of the patent.

In both cases, the license granted pursuant to these provisions:

- must not be exclusive;
- must not be assignable except in connection with the sale of a business;
- must be for a consideration agreed between the parties and if no agreement is reached, for a consideration determined by a court to be just and reasonable having regard to the economic value of the license; and
- is subject to any terms stated in the order.

Where an existing license is extended under these provisions, the court is entitled to take into account the terms and conditions of the previous license. The compulsory licensing provisions do not apply in the case of pharmaceutical patents where the term could have been extended under the repealed provisions.

Infringement of process patents

The amending Act imports the provisions of Article 34 of the TRIPs Agreement which provides that when infringement proceedings are commenced in relation to a patent for a process for making a product and the defendant alleges that the process used is different from the patented process, the court may determine that the product is made by the patented process, unless the defendant can provide evidence to the contrary. Such a determination is open to a court if it is satisfied that it is very likely that the defendant's product was made by the patented process and that the plaintiff has taken reasonable steps to find out the process actually used and has not been able to do so. The court is obliged to take proper account of the defendant's interests in having its trade secrets protected and is required to decide how a defendant can best adduce evidence to prove that its process does not infringe the patented process.

PLANT VARIETY PROTECTION LAWS

In 1987 the Federal Parliament passed the Plant Variety Rights Act, 1987 which conformed to the 1978 Act of the UPOV Convention. To bring the Australian law into conformity with the 1991 Act of the UPOV Convention the 1987 legislation was replaced by the Plant Breeders' Rights Act 1994.

Scope of plant breeders' rights

Generally, the plant breeders' rights (PBR) conferred by the Plant Breeders' Rights Act 1994 (henceforth called the Act) are defined in section 11 as:

> the exclusive right to do or to licence the following acts in relation to propagating material of the variety:
> (a) produce or reproduce the material;
> (b) condition the material for the purpose of propagation;
> (c) offer the material for sale;
> (d) sell the material;
> (e) import the material;
> (f) export the material;
> (g) stock the material for the purposes described in paragraph (a), (b), (c), (d), (e) or (f).

Excepted by section 16 from these rights are acts done privately and for non-commercial purposes, or for experimental purposes, or for the purpose of breeding other plant varieties. Seed saved by a farmer from harvested material and treated for the purpose of sowing a crop on that farmer's own land is considered by section 17 not to be an infringement. The section also provides for a particular taxon to be exempted by regulation. Section 18 provides that PBR are not infringed when propagating material is used as a food, food ingredient or fuel, or for any other purpose not leading to or involving the production or reproduction of propagating material. Finally, section 23 provides that PBR are exhausted following the sale of propagating material by a grantee unless there is a multiplication of the material after the sale.

Duration of plant breeders' rights

The general duration of PBR is provided by section 22 of the Act to be 25 years in the case of trees and vines and 20 years for any other variety. This duration commences from the date of grant of a PBR in the variety. Where a plant variety is declared under section 40 of the Act to be an 'essentially derived variety' from an initial variety, section 22 provides that the total duration of protection for the dependent or essentially derived variety can last for no longer than the duration of the protection of the initial variety.

Application for plant breeders' right

Eligible applicants

Section 24 of the Act states that a breeder can make application for a grant of PBR whether or not the breeder is an Australian citizen, or resident in Australia, or the variety was bred in Australia. The section provides for two or more breeders to make a joint application.

The right of a breeder of a plant variety to apply for PBR under the Act is declared by section 25 to be personal property and capable of assignment and of transmission by will or by operation of law.

Form of application

The form of application for PBR is prescribed by section 26. It provides that an application must contain: (i) the name and address of the applicant; (ii) the name and address of any agent, if any, making the application on the applicant's behalf; (iii) if the applicant is the breeder of the variety, a statement to that effect; (iv) if the applicant is not the breeder of the variety, details of the applicant's right to make the application; (v) a brief description, with a photograph, if appropriate, of a plant of the variety sufficient to establish a *prima facie* case that the variety is distinct from other varieties of common knowledge; (vi) the name, and any proposed synonym of the variety; (vii) particulars of the location at which, and the manner by which the variety was bred, including particulars of the names by which the variety is known and sold in Australia and particulars of any PBR granted in Australia or in another country which is a signatory to the UPOV Convention; (viii) particulars of any application for, or grants of, rights of any kind in the variety in any other country; (ix) the name of an approved person who will verify the particulars of the application, and who will supervise any test growing of the variety required under section 37 of the Act and who will verify a detailed description of the variety; and (x) such other particulars (if any) as are required by the approved form.

Application fee

An application fee may be prescribed under section 26(4) of the Act.

Acceptance or rejection

The Secretary of the Department of Agriculture (henceforth called 'the Secretary'), who is responsible for the administration of the Act, is required by section 30 of the Act to decide, as soon as practicable after an application is lodged, whether to accept or reject the application. Where the Secretary is satisfied that the application is prior in time to any other application and that it complies with the requirements of section 26 and establishes a *prima facie* case for treating the plant variety as distinct from other varieties, the application must be accepted. Upon acceptance the applicant must be notified that the application has been accepted and public notice of the acceptance of the application must also be given. Similar notification obligations apply where an application is rejected.

Variation of application

After an application for a PBR has been accepted, but before concluding the examination of that application, section 31 permits the Secretary to vary an application, subject to the payment of a prescribed fee. Section 32 requires the Secretary to notify the applicant for variation whether the request to vary has been accepted or rejected, setting out the reasons for the acceptance or rejection.

Withdrawal of application

An application is permitted by section 33 to be withdrawn by an applicant at any time. If this occurs after public acceptance of the application, the Secretary must, as soon as practicable, give public notice of the withdrawal.

Detailed description of the plant variety

As soon as practicable, but not later than 12 months after an application has been accepted, or within such further period granted by the Secretary, the applicant is required by section 34 to give the Secretary a detailed description of the plant variety to which the application relates. Failure to supply this description will result in the application being deemed to have been withdrawn. The detailed description must be in writing and in an approved form containing particulars of: (i) the characteristics that distinguish the variety from other plant varieties, the existence of which is a matter of common knowledge; (ii) any test growing carried out; (iii) any test growing outside Australia that tends to establish that the variety will, if grown in Australia, be distinct, uniform and stable; and (iv) such other particulars which may be prescribed. The Secretary is obliged by section 34 to give public notice of the detailed description as soon as practicable after it has been received.

Objection to application for PBR

A person may object under section 35 to an application for PBR if they can establish that their commercial interests would be affected by the grant of PBR to the applicant and that the Secretary cannot be satisfied that the various substantive requirements of the Act have been met by an applicant. The objection must set out the particulars of the manner in which the person considers his or her commercial interests would be affected and the reasons why the person considers that the Secretary cannot be satisfied that the various substantive requirements of the Act have been met.

Inspection of application and objections

Section 36 of the Act provides that a person may, at any reasonable time, inspect an application for PBR in a plant variety, or an objection lodged in respect of that application. Upon the payment of a prescribed fee, section 36 provides for a copy of an application or an objection to an application to be provided.

Test growing of plant varieties

In the case of an application for PBR which has been accepted, or an objection to such application, or a request for revocation of PBR, the Secretary may require a test growing, or further test growing, of the variety. In such case, section 37 requires notice to be provided to all relevant persons. The notice, in addition to telling the applicant, objector or grantee of the Secretary's decision, must specify the purpose of the test growing and may require the person to supply the Secretary with sufficient plants or propagating material and with any necessary information to permit the Secretary to arrange a test growing, or to make arrangements for an approved person to supervise the test growing and to be supplied with plants or propagating material. The expense of a test growing must be borne by the applicant, objector or person requesting revocation of the PBR. Section 38 provides for a test growing outside Australia of a plant variety which was bred outside Australia.

Provisional protection

Where an application for a PBR is accepted, the applicant is taken to be the grantee of that right from the date that the application is received until the application is disposed of. During this period of provisional protection, the applicant is prevented by section 39 from commencing any infringement action in respect of the PBR, until such time as the application is finally resolved in the applicant's favor.

Declarations of essential derivation

Where a person is the grantee of a PBR in a particular plant variety (the initial variety) and another person is the grantee of, or has applied for, PBR in another variety (the second variety) the grantee of PBR in the initial variety may seek a declaration from the Secretary under section 40 that the second variety is an essentially derived variety of the initial variety. A plant variety is defined in section 4 of the Act to be an essentially derived variety of another plant variety if: (i) it is predominantly derived from the other plant variety; (ii) it retains the essential characteristics that result from the genotype or combination of genotypes of that other variety; and (iii) it does not exhibit any important (as distinct from cosmetic) features that differentiate it from that other variety.

The application for essential derivation must be in an approved form and contain such information relevant to establishing a *prima facie* case of essential derivation. If the Secretary is satisfied or not satisfied as the case may be that a *prima facie* case has or has not been established, the applicant and the grantee of PBR in the second variety must be informed and provided an opportunity to rebut the *prima facie* case. Section 41 permits the Secretary to order a test growing in order to rebut a *prima facie* case of essential derivation. A similar test growing regime is provided for by the section to that contained in section 37.

Grant of PBR

Registrable plant varieties

Section 42 provides that PBR must not be granted to any variety of plant in a taxon declared by regulation to be one to which the Act does not apply. However, it is not envisaged that this provision will be implemented since the 1991 UPOV Convention requires that all plant varieties be eligible for PBR. A plant variety is considered to be registrable, pursuant to section 43, if the variety has a breeder, is distinct, uniform, stable and has not been or has only recently been exploited. For the purposes of this section a plant variety is distinct if it is clearly distinguishable from any other variety whose existence is a matter of common knowledge. It is uniform if, subject to the variation which may be expected from the particular features of its propagation, it is uniform in its relevant characteristics on propagation. A plant variety is stable if its relevant characteristics remain unchanged after repeated propagation. A plant variety is taken under section 43 not to have been exploited if it or propagating material has not been sold to another person by or with the consent of the breeder. For the purposes of this section, a plant variety is taken to have been only recently exploited if at the date of lodging the application for the PBR in the variety propagating or harvesting material has not been sold to another person by, or with the consent of, the breeder, in Australia, more than one year from that date. In the case of exploitation in other UPOV signatory states, the sale should not have been more than six years before that date in the case of trees or vines, or more than four years before that date in any other case. A plant variety is treated by section 43(8) as a variety of common knowledge where in addition to any other reason, an application for PBR in the variety has been lodged in a UPOV contracting state.

Grant of PBR

Where an application for PBR in a plant variety is accepted, section 44(1) provides that following examination of the application the Secretary must grant the right to the applicant where the Secretary is satisfied that: (i) there is such a variety; (ii) the variety is registrable within section 43; (iii) the applicant is entitled to make the application; (iv) the grant of that right is not prohibited by the Act; (v) the right has not been granted to another person; (vi) the name of the variety complies with section 27; (vii) propagating material of the variety has been deposited for storage, at the expense of the applicant, in a genetic resource center approved by the Secretary; (viii) in the case of a species indigenous to Australia a satisfactory specimen plant must be supplied to a prescribed herbarium; and (ix) all fees have been paid.

PBR is granted by the issue of a certificate in approved form. Section 45 provides that only one grant of PBR may be made under the Act in relation to a plant variety, irrespective of the number of owners of that variety, or whether that variety is an initial variety or a derived variety.

Effect of grant of PBR

If a person is granted PBR in a plant variety, section 48 provides for the grantee the right to take precedence over any other person who was entitled to make an application for the right in the variety. Such person is not prevented, however, from applying for a revocation of rights under section 50, or to seek administrative review of the Secretary's actions in relation to the grant of PBR, or to request the Secretary to make a declaration under section 39 that the right which was granted was essentially derived from another plant variety. Where it transpires that another person was entitled in law or equity to an assignment of the right to make an application for the PBR, that person is entitled to an assignment of the PBR.

Grant of PBR subject to conditions

Section 49 envisages that where the Minister for Agriculture considers it appropriate a PBR may be granted subject to conditions. In this regard, the Minister would probably take the advice of the Plant Breeders' Rights Advisory Committee, established under section 63 of the Act.

Revocation

Section 50 provides for the revocation of PBR or a declaration that a plant variety is essentially derived from another plant variety if the Secretary becomes satisfied that facts existed which, if known before the grant of the right or the making of the declaration, would have resulted in the refusal to grant the right or make the declaration. Revocation may also result from a failure to pay prescribed fees.

Within 7 days of the decision to revoke, the grantee or transferee of a PBR must be provided with particulars of the grounds of proposed revocation. That person then has 30 days to provide a written statement to the Secretary. Applications for revocation may be made by a person whose interests are affected by the grant of PBR in a plant variety or by a declaration of essential derivation. In the event of revocation or surrender of a PBR, section 51 provides for the particulars of revocation or surrender to be entered in the Register and to be published.

Compulsory licensing

Section 19 of the Act requires the grantee of PBR in a plant variety to take all reasonable steps to ensure reasonable public access to that plant variety. This requirement is taken to be satisfied if propagating material of reasonable quality is available to the public at reasonable prices, or as gifts to the public, in sufficient quantities to meet demand. For the purpose of ensuring reasonable public access, section 19(3) permits the Secretary to license an appropriate person to sell propagating material of plants of that variety, or to produce propagating material of plants of that variety for sale, 'during such period as the Secretary considers appropriate and on such terms and conditions (including the provision of reasonable remuneration to the grantee) as the Secretary considers would be granted by the grantee in the normal course of business.' An exception to the grant of a compulsory license applies in the case of a plant variety which has 'no direct use as a consumer product' (section 19(11)).

A person may make a written request to the Secretary under the section for the grant of a license where a person considers that a grantee is failing to ensure reasonable public access to a plant variety and that failure affects that person's interests. The request must set out particulars of the alleged failure and of the effect upon that person's interests. The Secretary is then required by section 19(6) to provide the grantee an opportunity within 30 days to satisfy him that the grantee is providing reasonable public access to a plant variety, or that he will comply within a reasonable time. Where the Secretary decides to grant a license under section 19, a public notice must be issued identifying the variety, detailing the particulars of the license which is proposed to be granted and an invitation to persons to apply for a license (section 19(8)). The Secretary is required to consider all applications and, at least one month prior to granting a license, must publicly notify the name of the proposed licensee, as well as notifying each of the applicants (section 19(9)).

Infringement of PBR

Infringing acts

Generally speaking, PBR in a plant variety is infringed by an unauthorized person: (a) doing one of the acts which are comprised in the PBR defined in section 11 of the Act; (b) claiming the right to do one of those acts; and (c) using the name of a registered variety in relation to another plant or another plant variety (section 53(1)). An action for infringement is brought in the Federal Court of Australia.

Defenses

An infringement will not occur where the act complained of is exempted from the operation of section 11, e.g. by sections 16–19 and 23. A defendant in an action for infringement may counterclaim for revocation of that right on the ground that the variety was not a new plant variety; or that facts exist which would have resulted in the refusal of the grant of that right. Under section 55 a person who proposes to perform an act described in section 11 may, by an action against a grantee of PBR in a plant variety, apply for a declaration that the performance of that act would not constitute an infringement of that right.

Remedies

The Court in an infringement action may grant an injunction subject to any terms which the Court thinks fit and, at the option of the plaintiff, either damages or an account of profits (section 56). Where a person satisfies the Court that at the time of the infringement he was not aware of, and had no reasonable grounds for suspecting the existence of that right, it may refuse to award damages or order an account of profits. This exoneration for innocent infringements is not available where propagating material of the plant variety, labeled so as to indicate that the PBR is held in Australia, has been 'sold to a substantial extent before the date of the infringement' (section 57(2)).

Administration

Registrar of Plant Breeders' Rights
Section 58 of the Act provides for the establishment of the Office of the Registrar of Plant Breeders' Rights, which is responsible for the general administration of the Act and for the maintenance of the Register of Plant Varieties.

Plant Varieties Journal
Under section 68 of the Act, the Secretary is require to issue a *Plant Varieties Journal* in which all public notices are to be published.

Genetic resource centers and herbaria
The Act in section 70 provides for the nomination of genetic resource centers for the storage and maintenance of germplasm material. An organization with the facilities for storing plant material may be declared a herbarium under section 71 of the Act.

Plant Breeders' Rights Advisory Committee

The Plant Breeders' Rights Advisory Committee is established by section 53 of the Act with the role of advising the Registrar on technical matters arising under the Act. Additionally, the Advisory Committee is required to advise the Minister on any regulations exempting taxes from the operation of the Act; any extension of the term of protection under section 22 of the Act.

Transitional

Plant variety rights granted under the Plant Variety Rights Act 1987 are preserved as PBR under section 82 of the Act and, under section 80. The Register of Plant Varieties under the old Act is incorporated into the Register established under section 58 of the Act.

REFERENCES

Blakeney, M. (1996a) *Trade Related Aspects of Intellectual Property Rights: A Concise Guide to the TRIPs Agreement.* Sweet & Maxwell, London.

Blakeney, M. (1996b) The impact of the TRIPs agreement in the Asia Pacific region. *European Intellectual Property Review* 18, 544–554.

Bureau of Industry Economics (1995) *The Economics of Intellectual Property Rights for Design.* Australian Government Publication Office, Canberra.

Copyright Convergence Group (1994) *Highways to Change: Copyright in the New Communications Environment.* Australian Government Publication Office, Canberra.

Industrial Property Advisory Committee (1984) *Patents, Innovation and Competition in Australia.* Australian Government Press, Canberra.

China

8

Tan Loke-Khoon

*Baker & McKenzie, 14th Floor, Hutchinson House,
10 Harcourt Road, Hong Kong*

INTRODUCTION

No other country has experienced such an unprecedented economic growth rate as the People's Republic of China (PRC) in the last ten years. Paramount Leader Deng Xiao Ping's economic reforms have changed the face of the PRC, bringing great opportunities for foreign investors and enterprising Chinese. Currently, one of the greatest concerns facing foreign investors is protection of their intellectual property rights (IPR). This chapter intends to give an easy to follow yet comprehensive guide to the laws and issues relating to intellectual property (IP) protection in the PRC.

On a world scale, the PRC has a relatively advanced framework of laws governing IP. It is a member of most international treaties governing IP and also a member of the World Intellectual Property Organization (WIPO), a United Nations (UN) Agency. Once China becomes a member of the World Trade Organization (WTO), it will also need to implement the Agreement on Trade-related Aspects of Intellectual Property Rights (TRIPs).

CURRENT STATUS OF INTELLECTUAL PROPERTY LAWS

Administrative framework

The PRC has set up a network of agencies and offices to administer IP at both national and local levels. These offices and agencies report directly to the State Administration for Industry and Commerce (SAIC), a government ministry under the State Council. Other branches of the government that participate in enforcement are the customs authorities, Public Security Bureau, the Procuratorate, Press and Publications Bureau and the Ministry of Culture.

For trademark and service mark matters, three offices under the SAIC administer registration and disputes: the Fair Trade Bureau, the Trademark Review and Adjudication Board and the Trademark Office. Administrations for Industry and Commerce (AICs) are established at the local levels in the PRC (provincial, city, county) and can be requested to conduct raid actions, after they have been presented with proof of the IPR and the infringing activities. The powers of the AICs are clearly specified in the PRC IP laws, giving the authorities the right to enter premises, seize accounts and records, seal and destroy goods and impose fines. The AIC is also empowered to award compensation after considering the severity of infringement.

For copyright matters, the National Copyright Administration (NCA) is in charge of the nationwide administration of copyrights, including copyrights belonging to foreigners. Like the AIC offices, there are also copyright offices at local levels in the PRC (usually the provincial and city levels). The PRC Copyright Law requires foreign complainants to file their complaints with the NCA at the central level in Beijing, regardless of where the infringement has taken place. For patent matters, the Patent Office is the government agency that handles legal and administrative matters.

The judiciary has established specialized IP courts to adjudicate IP cases. The judges who serve in these courts have been trained to handle IP cases. If an infringed party wishes to take action, in compliance with trademark, copyright and patent laws, the infringed party can take their case through the courts or through an administrative agency. In practice, administrative recourse has proved to be more effective because of the relative expertise of officials and their ability to conduct raid actions with little warning. Court proceedings are time-consuming and offer very little in the way of compensation. Additionally, a new Working Conference on Intellectual Property Rights was established as part of the 1995 China–US Agreement on Intellectual Property Rights.

Legislative framework

Trademarks and service marks

The laws governing trademarks and service marks are: the Trademark Law of the People's Republic of China (the 'Trademark Law'), which came into force on 1 March 1983 and was revised on 1 July 1993, and the Detailed Implementing Rules of the Trademark Law of the People's

Republic of China, which came into force in 1988 and has been revised twice since, in 1993 and 1995. The implementing regulations filled in some of the gray areas of the Trademark Law by clearly setting out procedures for trademark application, examination and registration processes. Amendments in 1993 expanded the scope of law to allow for companies to, *inter alia*, register service marks. Marks that possess a distinctive character may be registered. However, marks that are generic or descriptive of the quality, ingredients, functions or other characteristics of the products for which the trademark is to be used may not be registered. Once a trademark is registered, it is valid for ten years and may be renewed for consecutive periods of ten years. The PRC has adopted the 'first-to-file' rule for obtaining trademark rights, regardless of prior use. This means that the first applicant to file an application for registration of a mark will pre-empt all later applicants. There is, however, an exception to the rule. Under the Paris Convention for the Protection of Industrial Property, 'well-known' marks can be protected even if they have not been registered. For recognition of a well-known mark, Chinese authorities will generally require proof that the mark enjoys a leading market position in the original country of manufacture, enjoys a substantial degree of name recognition in the international market and has achieved a substantial degree of fame within the PRC. Although the Trademark Office has been willing on occasion to enforce the provisions of the Paris Convention relating to 'well-known' marks, it remains the case that trademark protection in the PRC is best secured through registration. In addition, in accordance with the Paris Convention, nationals of other member nations may claim priority use of trademarks within six months of the first filing.

Registration in the PRC is advisable even for companies that have not entered the Chinese market. PRC trademark registrations can be maintained by advertizing the marks once every three years in an approved PRC publication. Thus, companies that do not yet sell products or license trademarks in China due to investment restrictions can obtain trademark registrations and maintain their rights in anticipation of future use.

In addition, China has taken steps to bring its trademark protection laws in line with international standards. In 1988, the PRC changed their classification of goods to comply with the International Classification of Goods and Services, used by most trademark registries throughout the world. In October 1989, the PRC entered into the Madrid Agreement for the International Registration of Marks. The Madrid Agreement permits owners of so-called 'international registrations' to obtain trademark and service mark registrations in all other Madrid Union countries upon payment of a modest fee. An international registration is obtained by filing a trademark or service mark application in a country in which the applicant has a 'real and effective industrial commercial establishment'.

Copyright

The main laws governing copyrights in the PRC include: (i) the Copyright Law of the People's Republic of China (the 'Copyright Law') and the Implementing Regulations of the Copyright Law of the People's Republic of China, both effective from 1 June 1991; (ii) the Regulations for the Implementation of International Copyright Treaties, effective from 30 September 1992; and (iii) the Regulations for the Protection of Computer Software, effective from 1 October 1991.

The PRC extends protection to foreign works upon 'first publication' of a work in China, or within 30 days of publication elsewhere. As a result of the Memorandum of Understanding signed by the USA and China on 17 January 1992, all US works not in the public domain are now protected while works of residents of other member countries may be protected under relevant international conventions, for example, the Berne Convention for the Protection of Literary and Artistic Works (Berne Convention) and the Universal Copyright Convention.

Under the Copyright Law of the PRC, the following may be protected: literary works; oral works; musical, dramatic and choreographic works; works of fine art; photographic works; cinematographic works, television and video works; product engineering designs and their explanations; maps and schematic drawings and computer programs. Most copyrights are protected for the life of the author plus 50 years. In cases of copyrights originally vesting in a legal person, and copyrights in cinematographic, television and photographic works, and video and sound recordings, the duration of protection is 50 years from the date of first publication.

LICENSING AND COMPULSORY LICENSING OF COPYRIGHTS. Copyright holders can also license their rights for up to ten years (renewable upon expiry), after which the rights revert to the original owner. However, Articles 35, 37 and 40 contain language that permits a work to be used by any user who pays the owner fees fixed by the government. In this instance, permission need not be obtained from the copyright holder before using the copyrighted work. These articles permit compulsory licensing of any published work for commercial performances (Article 35), previously published works (including lyrics and composed works) for new sound recordings (Article 37) and previously published works for use in the creation of a new work by a television or radio station, excluding the broadcast of entire films or films for television (Article 40).

INTERNATIONAL COPYRIGHT TREATIES. China joined the Berne Convention in 1992. Rights protected in accordance with the Berne Convention include the exclusive right to make and authorize translations and reproductions, including visual and sound recordings of writings and the exclusive right to perform publicly or translate dramatic and musical works. In 1992, China also acceded to the Universal Copyright Convention which took effect in China on 30 October 1992. China also joined the Geneva Phonograms Convention with effect from 30 April 1993.

REGULATIONS FOR THE IMPLEMENTATION OF INTERNATIONAL COPYRIGHT TREATIES. Under the regulations, foreign works of applied art are protected for 25 years from their creation. Foreign computer programs are also protected as literary works that do not require registration, and are protected for 50 years from the end of the year of first publication. Also included in the regulations is the protection of foreign works that are created by compiling non-protectable materials that possess originality. There is also elimination of certain limitations imposed by the Copyright Law on the copyright owner's rights to comply with the Berne Convention.

Patents

In April 1985, the Patent Law of the People's Republic of China (the 'Patent Law') and its implementing regulations came into effect. This legislation is made up of features from patent laws drawn from a number of developed and socialist countries. The revised version of the Patent Law, together with the Detailed Implementing Rules for the Patent Law of the People's Republic of China became effective on 1 January 1993.

Effective from 1 January 1994, China became a member of the Patent Cooperation Treaty (PCT). Consequently the PRC Patent Office can now receive international applications filed by applicants in any contracting states of the PCT.

The Patent Law, like the Trademark Law, adopts a first-to-file system. Accordingly, the first inventor to file an application for an invention has the right to patents awarded with respect to the invention. Pursuant to the Paris Convention, however, if a patent application for an invention or utility model is first filed in another Convention-member country within 12 months before the filing date in the PRC, the prior filing date will be regarded as the priority date in the PRC. The relevant period is six months in the case of design patent applications.

Administrative protection of agrochemical products

Regulations for the Administrative Protection of Agrochemical Products were promulgated by the Ministry of Chemical Industry on 1 January 1993. This legislation provides for so-called 'pipeline protection' for agrochemical products patented between 1 January 1986 and 1 January 1993 in certain foreign countries and under stipulated conditions.

'Pipeline protection' is available to such products which are owned by individuals and enterprises from countries which have entered into bilateral agreements with the PRC (at present, these countries include the USA, Japan, Switzerland and members of the European Union).

The procedure for applying for 'pipeline protection' is set out in regulations issued by the Ministry of Chemical Industry (MCI). The regulation and implementing rules define agrochemical products as chemically synthesized agricultural chemicals, including herbicides, insecticides, fungicides, rodenticides and plant growth regulators produced by chemical synthesis. Enterprises, organizations and individuals in countries that have concluded a treaty or agreement with the PRC concerning administrative protection of agrochemical products may apply to the MCI for administrative protection.

To qualify for administrative protection, agrochemical products must satisfy the following criteria:

1. They must not have been eligible for protection under the PRC Patent Law prior to 1 January 1993.
2. They must have enjoyed exclusive rights through product patents granted in the applicant's home country between 1 January 1986 and 1 January 1993.
3. They must not have been marketed in the PRC prior to 1 January 1993.

Applications for administrative protection of agrochemical products must be processed through the China Zhengda Chemical Industry Legal Affairs Centre. Each application may cover only one agrochemical product. Applications must be submitted in written form without alterations and include the following in both Chinese and the official language of the applicant's country:

- the name, formula or prescription, and method of application of the product;
- a copy of the document issued in the applicant's country proving that the applicant has the exclusive rights to the product;
- a copy of the contract for the manufacture or sale of the product in China formally entered into between the applicant and a Chinese legal person; and
- a Power of Attorney in favor of the agent.

The term of administrative protection is for seven years from the date on which the certificate of administrative protection is issued. Foreign owners of such exclusive rights must pay an annual fee. Administrative protection may be terminated early if the exclusive rights to the product in the owner's country become void or if the owner does not apply to the State Council's administrative department of agriculture for permission to manufacture or sell its product.

If a protected agrochemical product is manufactured or sold without a license from the owner of the exclusive rights, the owner may request the Ministry of Chemical Industry to stop such activity and may institute an action in a People's Court for financial compensation.

Unfair competition and passing off

China's Unfair Competition Law was promulgated by the State Council on 2 September 1993. It provides some protection for unregistered trademarks, packaging, design and get-up. It also prohibits acts of unfair competition by monopolies or cartels to control prices in the market. Protection is also given to confidential information and business/trade secrets. The Unfair Competition Law prohibits business operators from engaging in the following acts of unfair competition:

- passing off of the registered trademark of another party;
- unauthorized use of the name, packaging or design peculiar to well-known packaging;
- unauthorized use of the enterprise name or personal name of another party; use of certification marks, marks of fame, marks of excellence that are counterfeit or used without authorization; and
- falsification of the place of origin or making of misleading and false statements as to the quality of the merchandise.

The Unfair Competition Law also prohibits the infringement of 'business secrets', defined as technical and business information that is 'private' and 'can bring economic benefits' to the rightful party and for which the rightful party has adopted measures to maintain its confidentiality. If a claimant wishes to take action against an infringer, the SAIC is empowered to impose fines and to order and supervise the return by the infringer of blueprints/drawings and software or other relevant materials.

Enforcement of intellectual property rights

To complement the existing IP laws, the PRC has a network of enforcement agencies which have been given the power to monitor and enforce compliance with the IP legislation. The administrative authorities, particularly the SAIC, have had recognized success in curbing IP infringements, largely due to an excellent network of offices throughout China. The NCA and the Patent Administration Bureau have been less active in this respect particularly due to lack of manpower, experience and resources.

IPR WORKING CONFERENCES. The State Council's Working Conference on Intellectual Property Rights consists of the heads of various central departments including the PRC Patent Office, the Ministry of Foreign Trade and Economic Cooperation (MOFTEC), the Ministry of Foreign

Affairs, the Ministry of Culture, the State Science Commission, the State Press and Publications Administrations, the Ministry of Justice, the SAIC, the PRC General Administration of Customs and the Ministry of Public Security. 'Sub-central IPR Conferences' will also be set up in at least 22 provinces and major cities. These will be under the direction of the State Council IPR Conference.

The main functions and duties of the Working Conferences are to coordinate and decide on the major policies and measures for the effective protection of IPR. To this end, the members of the conference will monitor the implementation of laws and regulations on IP; instruct and organize the relevant authorities within the regions to provide education and publicity for the laws regarding IPR; and improve law-enforcement skills of leading officials and enforcement personnel.

ENFORCEMENT TASK FORCES. Under the Working Conference system, the authorities with responsibility for IPR enforcement will participate in enforcement task forces (ETFs). Under the agreement, these authorities, including the SAIC, the NCA, the PRC Patent Office and the PRC Customs Authority, must provide the necessary resources and assistance to ETFs to ensure effective IPR enforcement. Each ETF should have the authority to initiate and carry out investigations of any suspected infringers, and carry out subsequent raid actions and seizure of infringing goods.

The major problems that ETFs face are mainly due to the lack of funding and comprehensive training of officials in IP issues. In some areas there are problems with local protectionism which adds to the difficulties of bringing infringers to task for their activities.

The judiciary

Most IP cases are handled by the civil and economic chambers of the Intermediate People's Courts, which occasionally suffer from a lack of trained personnel. Special IP chambers have been established in the larger provinces in China. At present, however, these chambers suffer from a lack of experience but could prove to be very important in the future for protecting IPR.

If a claimant wishes to take action against an infringer, the claimant must show evidence of the infringing products, infringing packaging, the identity and location of the infringing manufacturers and sellers and the location of the sales of the infringing product. Most foreign companies prefer to use the administrative route rather than the judicial route, as the former is less costly and usually more efficient in punishing the infringers.

THE WORLD TRADE ORGANIZATION AND CHINA

In 1947, the Nationalist government of China was one of the signatory nations to the General Agreement on Tariffs and Trade (GATT). After the Communist victory, the newly established PRC withdrew its membership, in 1950. Since 1986, the PRC has been trying to accede to the WTO, but as yet has not been successful. The main problems the PRC are facing are mainly due to what some countries perceive as unfair trading practices and the endemic weaknesses in the enforcement of IPR.

Before acceding to the WTO, the PRC must complete two different arenas of negotiations. In Geneva, a Working Party of members of the WTO is engaged in discussions with China on how to bring its domestic laws into compliance with WTO rules of fair trade, market-access IP, and other key areas. At the same time, China is holding bilateral talks with each member nation concerning tariff reduction and other market-opening issues. Only after both sets of negotiations are completed will China's bid come to a vote by the WTO's Ministerial Council. After ten years of negotiations, China feels it is time it is granted accession. There have been assurances from the USA that this will take place soon. The main negotiation issues include the following.

1. *Transition period.* China argues that because of its low per capita income, it should be treated as a developing country. This would allow it to take more time to implement WTO terms and be held to less rigorous standards than a developed country. Most WTO members disagree, due to the strength of the Chinese economy.

2. *Trading rights.* Beijing requires foreign companies to seek approval of each of their imports and exports. This practice contradicts WTO rules which allow foreign companies to import and export what they wish (with some exceptions). The negotiators would like China to change its 'trading rights' practices in approximately three years after accession to the WTO.

3. *Intellectual property.* Even though China has improved its system of protecting IP, counterfeiting of audiovisual products has led to estimated losses of billions of dollars to US software companies. The US–China IP Agreements of 1995 and 1996 have brought about marked changes. However, pirates are getting more sophisticated and there is still an overwhelming demand for counterfeit products, not only on the mainland but in other countries and regions including Russia and Hong Kong. Therefore, before the PRC accedes to the WTO, the PRC needs to assure members of the international community that they will continue in their efforts to curb IP abuses by pirates and other infringers of IPR.

Changes in legislation

There have been significant changes to the Copyright Law, due to increased international pressure. The US–China IP Agreement of 1995 and 1996 focused heavily on copyright infringement. As a result, China has called for an increased enforcement campaign against audio–video and computer software pirates, implementing new and stricter regulations of audiovisual production factories.

Customs officials have been granted greater powers to monitor exported products and imported machinery, particularly goods that can be used to manufacture audiovisual products. To monitor and trace audiovisual products, a title verification system for CD-ROMs, compact discs and laser discs has been implemented. By law, audiovisual products must also carry a unique identification code imprinted on the product surface.

TECHNOLOGY TRANSFER

The laws governing technology transfer have changed considerably since China opened its doors to foreign investment. China has encouraged the importation of foreign technology, but it has been criticized for not providing sufficient legal safeguards for protection of technology transfer. As such, it is still difficult for China to attract the latest technology. Foreign firms invest a lot of money and spend many years developing new products and, understandably, are concerned about the risk of entering the Chinese market.

There are two main laws governing technology transfers in the PRC: Regulations on the Administration of Technology Import Contracts, promulgated 24 May 1985 and the Detailed Rules for the Implementation of the Regulations on the Administration of Technology Import Contracts, promulgated 20 January 1988.

Under the Implementing Regulations, technology transfer is defined as:

> an assignment or license of patent rights or other industrial property;
> the provision of know-how such as production processes, formulas, product designs, quality-control, or management skills in the form of drawings, technical data and technical specifics;
> the provision of technical services.

Under these regulations, imported technology must be 'appropriate' to the economy of China. By being 'appropriate', the technology should be 'capable of developing and producing new products'. The authorities are selective about what type of technology they encourage. China is particularly interested in the following industries: communication, energy, necessary raw materials and machinery for the mechanical and electronics industries.

Technology import contracts must be approved by MOFTEC, one of the ministries under the State Council or one of its local delegates. MOFTEC is also responsible for approving investment contracts.

Generally, a Foreign Investment Enterprise (FIE) may import machinery and technology that is needed for production provided that the resulting products are exported. If the resulting products are intended to be sold domestically, a license is required to import machinery and technology.

MOFTEC also issues most import/export licenses. A license application will first go to MOFTEC, who will then refer the application to the State Planning Commission (SPC), which will then coordinate with the relevant ministry as well as the Economic Trade Commission of the State Council. All licenses are subject to quotas determined by the SPC, who also stipulate guidelines of commodities which are needed in the PRC.

CONCLUSION

The IP laws of the PRC, and the courts and administrative agencies that enforce them, provide a strong framework for the protection of IP in the PRC. The greatest challenge facing the PRC is effective and consistent enforcement of this framework. Local protectionism and lack of trained personnel contribute to the weaknesses in enforcing IP protection. The PRC needs to continue on its course of training enforcement personnel and judges, so that it can develop into a country where both foreign investors and local interests can be confident that their IPR will be fully and effectively protected. A safer country for IP holders will attract increased foreign investment, particularly in more advanced technologies, which would in turn assist China in its path towards modernization and long-term economic growth.

Japan

9

Kazuo N. Watanabe,[1] Atushi Komamine[2] and Yoshihiko Nishizawa[3]

[1]*Department of Biotechnological Science, Institute of Bioscience and Technology, Kinki University, 930 Nishi-Mitani, Naka-Gunn, Uchita, Wakayama 649-64, Japan;* [2]*Faculty of Science, Nihon Women's University, 2-8-1 Mejiro-dai, Bunkyo-ku, Tokyo 112, Japan;* [3]*Sumitomo Chemical Industry Co. Ltd, 818-1 Hourensan Soenishi-cho, Nara-city, 630 Nara, Japan*

NATIONAL PERSPECTIVE

Current status of intellectual property laws

General legal status

Intellectual property rights (IPR) laws are well established in Japan (Table 9.1; Anonymous, 1996a). Japanese IPR principles are based on the first-to-file system, which is different from the USA's first-to-invent system. The patent examination standards were revised in 1993 with an eye towards international harmonization. The revisions emphasize: (i) simplification of standards by integration of previously vague rules; (ii) enforcement of inventors' and owners' rights by extending the covered subjects/categories in a patent application; and (iii) adding standards in computer software and biotechnology. The general procedures on the patent application and granting are shown in Figure 9.1 (Anonymous, 1996a). Note, particularly, from this figure that:

1. The examination process is waived for the utility model as of 1 January 1994.

2. When an individual employee creates publication/presentation materials for an organization, copyrights belong principally to the organization.

3. In contrast, industrial property rights can be owned by individual employees, even when the invention was made while the employee was working for the employer. However, the employer could use the industrial rights without compensating the employee.

4. Since December 1990, on-line application via the Internet has been available.

Table 9.1. Intellectual property rights in Japan.

Category	Type of rights	Law	Subject protected	Key points for granting protection	Protected period
Industrial property rights	patent	patent law	invention	availability for industry; innovativeness; improvement	20 years after application
	utility model	utility model law	idea/device	availability for industry; innovativeness; improvement	6 years after application
	registered design	registered design law	design	availability for industry; innovativeness; creativity	15 years after registration
	registered trademarks	registered trademarks law	trademark/brand/ service mark	distinguishable from others	10 years after registration (renewable)
Copyrights	copyright property	copyright law	publications	originality	50 years after publication
	neighboring right	copyright law	all rights associated with publications	originality	50 years after publication or 50 years after death of author/performer
	moral right	copyright law	all rights associated with author/performer	originality	no fixed term
Others	trade secret	unfair competition prevention law	industrial rights	—	conditional with associated right
	protection of allotment of semiconductor circuit	law for protection of allotment of semiconductor circuit	idea/device on allotment of semiconductor circuit	originality	10 years after registration
	crop variety protection	seed and seedling law	plant varieties	originality; improvement	15 years after registration 18 years for perennials

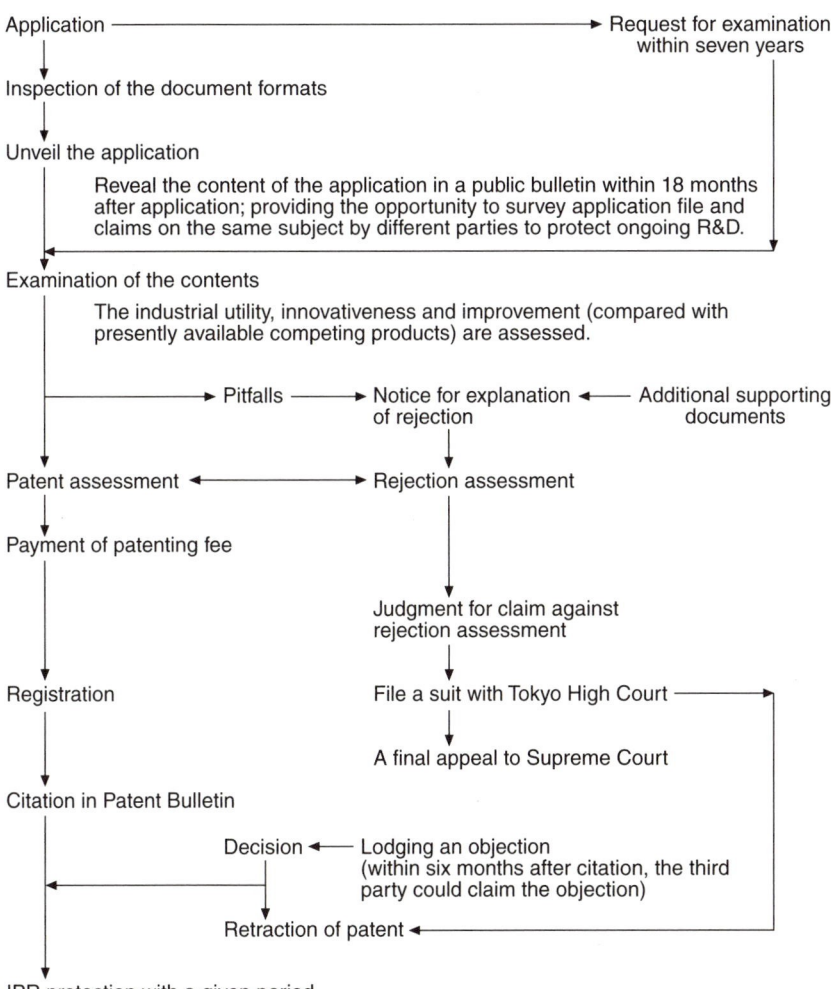

Fig. 9.1. A flow chart showing the procedure for patent application and appeal.

5. Since September 1994, commercial insurance premiums are available for covering legal actions against infringement.

The laws listed in Table 9.1 are internationally harmonized. Japan has participated in the World Intellectual Property Organization (WIPO) since April 1975. While associated with WIPO, Japan also participates in the following treaties:

- Paris Convention for the Protection of Industrial Property since July, 1899;
- Berne Convention for the Protection of Literary and Artistic Works in 1899;

- Universal Copyright Convention of 1952 under UNESCO since 1956;
- Patent Cooperation Treaty (PCT) since October, 1978;
- Union for the Protection of New Varieties of Plants (UPOV) since September, 1982 (this was based on the International Convention for the Protection of New Varieties of Plants, etc.).

Since 1956, Japan has also had a branch office of the Association Internationale pour la Protection de la Propriété Industrielle (AIPPI) and now has her own company called AIPPI-JAPAN. Japan does not participate in the treaties of Hague (1925), Rocarno (1968), Vienna (1973) or Lisbon (1958). In addition, there are many international disputes and imbalances, such as in biotechnology which is discussed later in this chapter.

Special attorney for IPR: Benri-Shi
It is very difficult to obtain a permit to become a Benri-Shi – a Japanese patent attorney. On average, only 3% of applicants pass the examinations. There is no systematic way to train people for the examinations, and only a limited number of private schools offer know-how courses on how to pass the examinations. Furthermore, there is no professional training system through which one can become an effective Benri-Shi. In addition to acting as an attorney, the Benri-Shi is required to be specialized in particular subjects, to a level equivalent to a PhD degree, so it is really challenging to acquire this legal permit and to give competitive client services. In the private sector, each company has specialized people for preparing the application before it is given to the patent attorney. Consequently, many researchers with science and technology PhDs are encouraged by their employers to obtain the patent attorney license so that the company does not depend on an external patent attorney.

Intellectual property rights at academic institutions
Intellectual properties (IPs) are not systematically managed and protected in academic institutions. Unlike US academic institutions, Japanese universities merely have their own office or research foundation which takes care of the various issues associated with proprietary materials and technology transfer. During protection and negotiation of IPs, the lack of an IPR administrative office causes complications for the individual researcher. Also, in case of licensing and settlement of royalty fees, individual owners face various problems since only partial support is available from the university administration. It is illegal for employees of public institutions to negotiate while at work. The government may compromise in the near future. Often these inventors give all their rights to the private sector, so in return, their research programs and their universities will receive compensation. A similar situation applies to central and local governmental agencies.

Relationship between intellectual property rights and agriculture

Plant variety protection laws
Since 1982, Japan has been a member of UPOV and follows its Revision of March 1991. While new plant varieties, based on agricultural biotechnology, could be protected by two new major IPR and seed/seedling laws, double protection is not allowed. Also, Japan is far from achieving a consensus and from making the necessary legal rules in the areas of newly emerging agricultural biotechnology and variety development.

Seed and seedling laws: aegis or Damocletian sword?
Dual controls in agro-biotechnology are associated with plant cultivars. Registration of a new variety is easy, since little information is given to the legal body, but production of a certified propagules is another issue, because for many crop species it is regulated by the government. This dual control structure on the registration and propagule production makes it difficult for the private sector to make a profit in crop species such as potatoes.

In evaluating the pros and cons of seed and seedling law, it is helpful to discuss the (i) protection of IPR and/or breeders' rights on cultivars; (ii) discouragement of seed/seedling business in the private sector due to the strict governmental control; (iii) quarantine enforcement by the government; and (iv) protection of domestic but weak seed/seedling industry against strong competitors.

Biodiversity
Since 1980, Japan has been a member of the Washington Convention (Convention on International Trade in Endangered Species of Wild Fauna and Flora, or CITES). Japan is also a member of the Convention on Biological Diversity (CBD) since December, 1993, and is considering incorporating the revised CBD on biotechnological aspects, particularly on the genetically modified organisms, which will come into effect in 1998. Japan also made its own law on the conservation of endangered species, which became effective in April 1994. Although international trade in the endangered species has been prohibited in Japan since CITES came into effect, domestic trade in endangered species within Japan was permitted until 1995. With the 1995 revisions of the law dealing with conservation of endangered species, trade in endangered species is now strictly regulated.

In order to conserve biodiversity and to ensure sustainable utilization, the Japanese government put together a national biodiversity strategy in October 1995. Features of the national strategy include:

- biodiversity in Japan and global aspects;
- fundamental national indices and long-term perspectives on conservation of biodiversity and sustainable utilization;

- actions and planning for implementation;
- interdisciplinary approach among central and local governments, private sector and individual Japanese citizens; and
- continuous self-examination and frequent revisions.

One specific problem associated with biological diversity in Japan is the poor recognition of IPR in genetic resources in the private sector, academic institutions and individual citizens, compared with that in other industrialized countries.

Many private Japanese companies associated with agro-biotechnology conduct their own 'secret' exploitation of plant genetic resources in many countries rich in such resources. These rights have been claimed retrospectively and/or are in the process of litigation in the countries of origin. Japanese private firms do not acquire and transfer germplasm by official routes. In many cases, these problems have been caused by improper brokers, and also by the lack of recognition of the international treaties and movements.

On an individual basis, there are a number of unsophisticated travelers aiming for profit who lack knowledge of IPR issues on plant genetic resources, and who consequently smuggle germplasm. This causes plant quarantine and IPR problems. Similar tendencies can be observed even in germplasm laboratories at Japanese academic institutions. Programs of education on PGR are getting weaker due to retirement of experts from these institutions.

Finally, Japan lacks integration between international and national germplasm repositories. In some cases, the ministries of Agriculture, Forestry and Fishery, and of Education and Culture perform duplicate functions. More specifically, the policing of germplasm acquisition and transfer is not well organized.

Pitfalls in intellectual property rights laws

The Japanese IPR laws are under the supervision of the Patent Agency and the Agency for Cultural Affairs. However, the Seed and Seedling Law is under the Ministry of Agriculture, Forestry and Fishery. Additionally, several ministries are associated with various IPR, including the Ministry of International Trade and Industry (MITI), Science and Technology Agency (STA), Ministry of Health and Welfare, etc. These agencies are influencial in determining grants or implementing IPR. The interaction between agencies and the harmonization of their systems are poor compared with those in other developed countries. Thus, with respect to agro-biotechnology, two or more agencies can give contradictory rulings. This makes the system very inconvenient for applicants and potential users of IPR. A good example concerns the biosafety and food safety rules governing genetically engineered organisms. Although five governmental agencies regulate this area, their rulings are inconsistent. Unless a supervisory group is established to watch over these agencies and integrate them, Japanese IPR laws, particularly those associated with biotechnology, may have a fragile base.

COMMERCIALIZATION, NATIONAL LINKS AND TECHNOLOGY TRANSFER TO THIRD PARTIES

Major issues in Japanese bioindustry

Biotechnology in Japan is not merely used to develop new production methods for substances. It is also extensively used to develop new pharmaceuticals, agrochemicals, seeds and seedlings, livestock, fish, and new reactions based on engineered enzymes.

Although Japanese bioindustry is very large with many assets, and is growing rapidly, there are many factors which suggest that the future may not be very bright. One of the main issues which will confront Japanese bioindustry in the future is to identify uncertain economic and social elements and implement measures to respond properly to them. The first issue is research funding for biotechnology. Biotechnology research requires an integration of various proprietary materials from molecular biology, molecular genetics, plant physiology, plant breeding and plant cell biology. An interdisciplinary approach must be implemented. Furthermore, it should be emphasized that biotechnology research requires longer periods than other research. Normally in Japan, ten years is needed to develop pharmaceuticals and plant-related products. This starts from the first confirmation of laboratory research results and concludes with the marketing of a final product.

The second major problem facing Japanese bioindustry is how IPs can be protected. Biotechnology covers a wide range of research, both basic and applied, as well as development and profitable product production. It is increasingly difficult for a private company to monopolize all possible patents for a single product. Therefore, the possibility of patent problems ending up in court is increasing. Once litigation starts, enormous amounts of time and money can be spent without reward. Also, because biotechnology is a relatively new area, there are very few legal precedents.

Alternatives to litigation in Japan include arbitration and cross-licensing. Because double protection has been eliminated throughout the revision of the UPOV Convention, it remains unclear as to how patent laws and seed and seedling laws will be applied.

Proprietary biotechnologies have been transferred to developing countries, especially to neighboring Asian countries, through the ODA basis (Kainuma, 1995; Takase, 1994), by private sector investment (Sumida and Nishizawa, 1995) and through independent initiatives (Kozai *et al.*, 1993; Altman and Watanabe, 1995; Watanabe and Pehu, 1997). The key pitfalls of proprietary technology transfer are: (i) scarce human resources in facilitating the entire technology transfer activities; (ii) lack of interest of philanthropic technology transfer in the private sector; and (iii) weak understanding of public agencies (Watanabe and Raman, 1997).

Safety issues hamper the recognition of proprietary technologies and products
Safety issues and public acceptance of genetically modified organisms (GMOs) are an increasing concern. These two major aspects of regulatory issues and their associated public concern are discussed below.

SAFETY ASPECTS. There are two schools of thought related to safety evaluation in biotechnology. The first takes a process-oriented approach. Safety must be evaluated wherever biotechnology-based processes are used in any product, regardless of the type of final product – be it a crop or live vaccine. Germany is addressing safety issues in this manner. In other words, any product that is manufactured using a biotechnology process should be regulated equally, no matter what final form it takes. The second school of thought is a product-oriented approach. This approach requires that, whether or not the biotechnology is used, each product should be evaluated based on scientific knowledge.

THE PUBLIC CONCERN. If the public is not well enough informed, they will not have the opportunity to choose whether or not to use the products of biotechnology (Zechendorf, 1994). Japan's public enlightenment and further education programs are weak (Macer, 1994, 1997). Today, consumers' associations and concerned individuals are gradually awakening to the introduction of agro-biotechnology products. These products are mainly from the USA. As of January 1997, products derived from seven North American transgenic cultivars have been deregulated for importation.

Soybean in particular, of which 95% is imported, has caused a major dispute between distributors, cooperatives and consumers. These groups cannot decide how to distinguish biotechnology-derived soybean products from non-biotechnology-derived products. This dispute continues because soybean products form a major part of the Japanese diet. Soy sauce, soybean paste, tofu and soybean curd are consumed every day by most people. Now is the time for both public and government agencies to discuss how agro-biotechnology products can be recognized and used, and whether the products should be accepted or banned.

The agro-biotechnology industry in Japan

In contrast with North American and European biotechnology associations, there is insufficient legal IPR protection and licensing negotiation by public industry associations. As previously noted, there are no biotechnology specialists among Japanese patent attorneys, although there are more than 500 patent attorney offices available. A facilitator or business consultant is needed to obtain income by enhancing licensing and product development from proprietary agro-biotechnology.

Overall, Japan can be regarded as a silent player in the agricultural biotechnology sector (Okada, 1996). Japan's private sector has lost many opportunities because of its rather cautious approach. The key patents in plant biotechnology have been taken by North American and European private companies (Stone, 1995), which has prevented Japanese companies from moving into profitable materials. The Japanese bioindustry faces still more obstacles associated with the international aspects of IPR, especially in North America (Hoyle, 1996). These include: (i) complication of patenting inspection (O'Shaughnessy, 1996); (ii) tendency to grant wider coverage of patentable subjects such as DNA sequences (Agris, 1996); and (iii) further changes in laws on patentable 'processes' (Van Horn and Barlow, 1996).

Japan is now learning from elsewhere, especially from North America and European countries. A strong leadership with long-term strategies and patience will form and guide the overall national alliance. It is only now been gradually proven that newly emerging companies are making progress by getting into unconventional markets with new products.

Eventual integration of the private sector in agro-biotechnology is shown by the following examples.

Mitsubishi Chemical is the seventh largest revenue-generating company in the world, and was the product of Mitsubishi Kasei and Mitsubishi Petroleum (Anonymous, 1996b, c). Also, Mitsui Petroleum Chemical is planning to merge with Mitsui Toatsu Chemical, to form Mitsui Chemical. Other biotechnology initiatives in the private sector also have domestic and international merger and acquisition (M&A) such as flower biotechnology business made by Agri-biotechnology Division of Kirin Brewery Co. (Okamura and Kagami, 1997). Therefore, by increasing their capital, assets, infrastructure, technology, market network and human resources, these Japanese private companies are rejuvenating and promoting their international competitiveness. Cross-licensing of proprietary biotechnology continues between Japanese biotechnology companies and international agro-biotechnology giants such as Monsanto. The Japanese government has also increased aid in overall R&D and emphasized several areas including biotechnology. This will give both the private and public sectors a chance to strike back into international markets with their own proprietary technologies.

ACKNOWLEDGEMENTS

This chapter was developed with the assistance of the International Agri-Biotechnology Association of Japan (IABA-Japan). K.W. was supported partly by the Research for the Future (RFTF) Program under the Japan Society for the Promotion of Science, project no. JSPS-RFTF96L00603. A.K. also acknowledges the RFTF Program for their support. Y.N. thanks the Japan Bioindustry Association.

REFERENCES

Agris, C.H. (1996) Prior art consideration when patenting DNA sequences. *Nature Biotechnology* 14, 1309–1310.

Altman, D.W. and Watanabe, K.N. (eds) (1995) *Plant Biotechnology Transfer to Developing Countries.* R.G. Landes Co., Austin, Texas, 300 pp.

Anonymous (1996a) *Brochure on Intellectual Property Rights.* AIPPI-Japan, Tokyo.

Anonymous (1996b) Association of German biotech companies launched. *Nature Biotechnology* 14, 1527.

Anonymous (1996c) A chemical decomposition. *Economist* November 6, 1996, 78.

Hoyle, R. (1996) Another salvo in the plant biotech wars. *Nature Biotechnology* 14, 680–682.

Kainuma, K. (1995) The role of JIRCAS in international technology transfer related to biotechnology application to agriculture and food processing in Japan. In: Altman, D.W. and Watanabe, K.N. (eds) *Plant Biotechnology Transfer to Developing Countries.* R.G. Landes Co., Austin, Texas, pp. 225–233.

Kozai, T., Uchimiya, H., Ishikawa, F. and Komamine, A. (1993) Role of agro-biotechnology for the conservation of global environment. *Nogyo and Engei* 68(10), 51–56.

Macer, D.R.J. (1994) Bioethics may transform public policy in Japan. *Politics and Life Sciences* February, 89–90.

Macer, D.R.J. (1997) Major concerns on plant biotechnology applications in plants: safety issues and bioethics. In: Watanabe, K.N. and Pehu, E. (eds) *Plant Biotechnology and Plant Genetic Resources for Sustainability and Productivity.* R.G. Landes Co., Austin, Texas, pp 87–101.

Okada, M. (1996) Japan/US comparisons of biotechnology patents. *Eubios Journal of Asian and International Bioethics* 6(6), 166–168.

Okamura, M. and Kagami, Y. (1997) Flower production in Japan and global strategies on agribio-technology and business. In: Watanabe, K.N. and Pehu, E. (eds) *Plant Biotechnology and Plant Genetic Resources for Sustainability and Productivity.* R.G. Landes Co., Austin, Texas (in press).

O'Shaughnessy, B.P. (1996) Patent pitfalls among the unpredictable arts. *Nature Biotechnology* 14, 1028–1029.

Stone, R. (1995) Sweeping patents put biotech companies on the warpath. *Science* 268, 656–658.

Sumida, S. and Nishizawa, Y. (1995) R&D cooperation in biotechnology with development countries. In: Altman, D.W. and Watanabe, K.N. (eds) *Plant Biotechnology Transfer to Developing Countries*. R.G. Landes Co., Austin, Texas, pp. 279–287.

Takase, K. (1994) *Evolution of Japan's Strategy on Global Agricultural Development*. IDCJ Study Summary Series No. 4. International Development Center of Japan, Tokyo, 40 pp.

Van Horn, C.E. and Barlow S.A. (1996) The new section 103(b) of the patent law: an obvious solution. *Nature Biotechnology* 14, 773–774.

Watanabe, K.N. and Pehu, E. (eds) (1997) *Plant Biotechnology and Plant Genetic Resources for Sustainability and Productivity*. R.G. Landes Co., Austin, Texas, 247 pp.

Watanabe, K.N. and Raman, K.V. (1997) Plant biotechnology and plant genetic resources: a global perspective. In: Watanabe, K.N. and Pehu, E. (eds) *Plant Biotechnology and Plant Genetic Resources for Sustainability and Productivity*. R.G. Landes Co., Austin, Texas, pp. 1–13.

Zechendorf, B. (1994) What the public thinks about biotechnology. *Bio/Technology* 12, 870–875.

India

10

Prabuddha Ganguli

Corporate Planning, Hindustan Lever Ltd, Hindustan Lever House, 165/166 Backbay Reclamation, Mumbai 400 020, India

INTRODUCTION

Science and technology in India, and intellectual property rights – looking back

The history of science and technology in the Indian subcontinent can be traced back to the Palaeolithic era (150,000 years ago) when a range of stone tools were in existence. This was followed by the flourishing period of Harappan culture (2500–1750 BC) during which beginnings were made in exquisite pottery, metal working, glazing, animal husbandry, agricultural implements and practices. Commercial trade links were also established with the neighboring cultures in the Central and Western Asian regions. The subsequent vedic period (1500 BC to 1200 AD) saw rapid advances in the understanding of biological sciences, astronomy, mathematics, various forms of materials, medicine including Ayurveda and surgical techniques, agricultural practices and evolution of technological skills specifically with respect to working with metals, ornamental pottery and tiles (Bose *et al.*, 1971). The scientific activity was essentially individualistic in pursuit of the ultimate truth. There was no competitive overtone and the knowledge was passed on through the strong family systems in the community.

In the next few centuries, the social milieu, traditional compulsions, political vicissitudes and caste distinctions did not promote cross-fertilization of disciplines or transfer of knowledge across family/caste borders. Knowledge and techniques that had been developed were retained within families and this was considered adequate protection to retain uniqueness. 'Inbreeding' of thought processes within a select class of people affected the creative spirit; the deleterious effects of this were apparent between 1200 and the mid-nineteenth century.

From the sixteenth and seventeenth centuries, the Portuguese, Dutch, French and British established commercial and colonial interests in India (Bose *et al.*, 1971). In order to exploit the natural resources and local talent to the full, the British introduced modern scientific methods and education in India, and India maintained a competitive commercial edge in trade. As competition from other European countries grew, the British introduced an Act of Protection of Inventions, based on the British Patent Law of 1852, which was enacted in India in 1856. By this Act, certain privileges were granted to the inventor for new methods of manufacture. Later changes to the law in this field were the Patents and Designs Protection Act 1872 and the Protection Inventions Act, which was introduced in 1883 and consolidated as the Invention and Designs Act in 1888 (Narayan, 1985).

Indian patents and designs came under the management of the Controller of Patents and Designs on 15 August, 1947. After independence, this Act was nationalized (Chand, 1950; Ayyangar, 1959). A Patents Bill was introduced into Parliament in 1965, but did not go through, so an amended Bill was introduced in 1967, resulting in the Patents Act of 1970, which, together with the rules, came into force on 20 April 1972. This Act is known as the Indian Patents Act 1970 and is still in force.

The Designs Act was brought into India in 1872 to extend protection to textiles, linen, cotton, calicoes and muslin; this included patterns/prints and modeling, casting, embossment of ornaments or articles of manufacture. The Indian Patent and Designs Act came into force in 1911 with amendments in 1978 (summarized in Appendix 10.1 at the end of this chapter) with the rules amended in 1985. These are currently in force.

Legislation to protect trademarks came into force on 6 June 1942 and was based on the principles of English Common Law . The Act of 1940 was further amended to the Indian Trade and Merchandise Marks Act 1958 which came into force on 25 November 1959. This act is still in force.

The Ayyangar report recommending changes in the Patent Act 1911 was an outcome of an analysis of the status of expertise in industry, trade, science and technology in India just after independence. The Patent Act of India 1911 was fairly liberal as patenting of products related to foods, pharmaceutical, chemicals, etc. was available with a full term of 16 years. This was directly in line with the British Patent Act of 1907. The Indian Patent and Designs Act of 1970 brought in significant changes with a number of restrictions related to patenting of inventions, specially in the area of chemicals, pharmaceuticals, agro-chemicals, foods, in which product patents had been discontinued and patenting of processes with a restricted life of seven years from the date of filing of the complete specification (or five years from the date of sealing the patent, whichever is shorter) was introduced. This protected patent regime provided a safe platform on which pharmaceutical and chemical industries could strike roots and grow in India and also meet the need to increase production rather than relying on imports, which was critical for the national economy. For example, pesticide imports were reduced from around 12,000 tonnes in 1965–66 to a mere 1100 tonnes in 1992–93. A number of new processes and technologies for production of pesticides have been developed by the Indian National Laboratories and transferred to Indian industries. Few efforts have been directed to inventing new molecules/products; instead effort has focused on development of alternative cost-effective manufacturing processes.

The highly protective industrial policies in India, coupled with a strong internalized and protective patent regime, resulted in lack of concerted efforts and development of expertise in search of new molecules or products in chemicals, pharmaceuticals, speciality materials and biotechnology. The restrictions summarized under sections 3, 4 and 5 of the Patent Act also led to lack of research directed towards new speciality materials including semiconductors, smart materials, alloys, optical glasses, etc., as these fell within the 'non-patentable materials' category in India. The industries in their turn considered that it was cheaper and simpler to collaborate and purchase technology rather than to initiate their own R&D programmes in India. This led to a rapid development of industry, with significant short-term benefits. However, these materials and technologies have advanced rapidly at the international level, becoming highly specialized, and Indian industry has lagged behind because it did not invest sufficiently in R&D to develop a critical R&D mass and keep its technology up to date. The recent opening up of the economy poses major challenges to Indian industries as they will have to re-adjust their strategies for survival in view of the growing competition in India from international corporations.

IPR ADMINISTRATION IN INDIA – CURRENT PRACTICES

The four branches of IPR are divided into two categories – (i) those which have a major role in industry, namely patents, designs and trade-marks, fall under the Ministry of Industries and are controlled by the Controller General of Patents, Designs and Trademarks (CGPDTM) and (ii) those which are considered to have cultural aspects, namely copyright, are part of the Ministry of Human Resource Development.

Patents are filed at one of the branches of the patent office in the four regions, which have distinct jurisdictions as shown in Figure 10.1. The inventors or assignees are expected to file the patent application in the regional patent office branch within whose jurisdiction they come. For example, if an institution has a registered office in Bombay, it should file its patent application at the Bombay Branch. The Council of Scientific and Industrial Research (CSIR) has 40 national laboratories in India located in various parts of the country, but patent applications from any of the laboratories within its control must be filed at the Delhi Branch of the Patent Office as CSIR is a registered society in Delhi. Foreigners who do not have a registered office in India have their patent applications filed through registered patent agents in India, and the office at which the application is to be filed depends on the agent's registered address.

The Patent Head Office in Calcutta has specific functions related to granting and sealing of patents, amendment of granted patents, restoration of lapsed patents, registration of assignments, transmission, etc., making entries in the register of patents, loss or destruction of patents, providing information related to patents, revocation of patent of amendment of complete specification on directions from the central government, etc.

Opposition-related activities primarily fall within the jurisdiction of the branches of the patent office. The patent applications are prosecuted

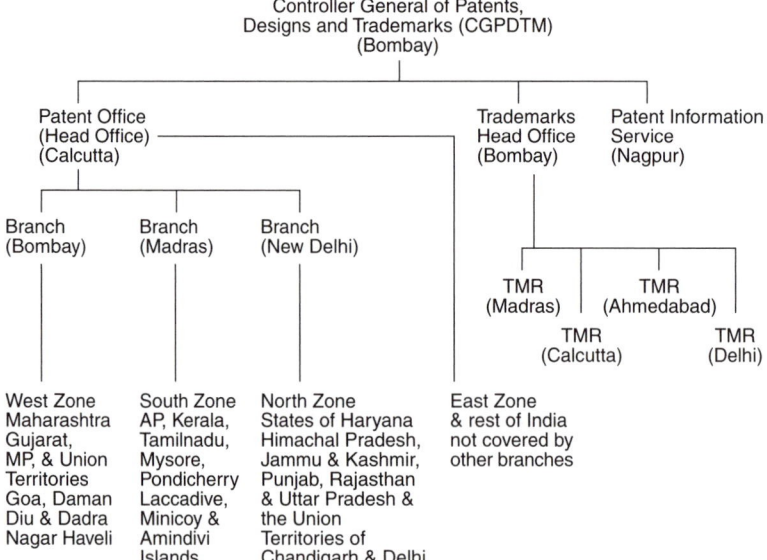

Fig. 10.1. Patents, designs and trademark administration in India. The addresses of the head office and branch offices of the Patent Office are given in Appendix 10.3, together with the addresses of other patent-related organizations.

in the branch office where the patent application is filed. Similarly, patent opposition filings and proceedings are carried out and completed in the branch office where the patent application was originally filed and prosecuted. The matter can move to the High Courts only if the parties involved are not satisfied with the outcome of the opposition proceedings held by the controller at the branch of the patents office.

Enforcement of patents and designs (infringement and revocation) in India falls under the jurisdiction of the High Courts. Compensations and injunctions on such matters are also decided by the High Courts. The patent head office in Calcutta only acts on the orders passed by the High Courts. Amendments are under way to simplify the steps of appeal.

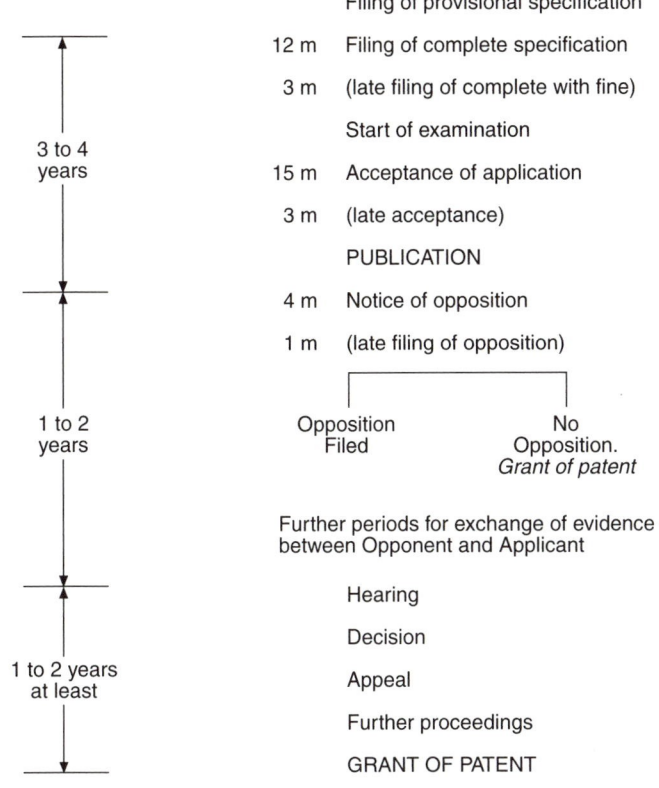

Fig. 10.2. Timescale involved in obtaining a patent in India. Note that: (i) examination of patents on inventions related to processes/methods of manufacture of drugs/foods etc. are taken up within 6 months of submitting the complete specification; (ii) patent in India is valid for 14 years from the date of filing the complete specification. For processes in the food and drugs sectors it is seven years after filing the complete specification or five years from the date of sealing the patent, whichever period is shorter.

Indian Patent Act 1970

To appreciate the role patents have played in India so far and the possible changes in the new patent regime as a result of India becoming a signatory to General Agreement on Tariffs and Trade (GATT), it is useful to understand some of the unique features of the Indian Patent Act 1970, which is currently in force. The aspects which are generally common to the patent laws in other countries are not elaborated here. India follows the 'first-to-file' system as in most countries.

Patentable inventions
'Invention' is defined by the Indian Patents Act 1970 as any new and useful: (i) art, process, method or manner of manufacture; (ii) machines, apparatus or other article; (iii) substances produced by manufacture, including any new and useful improvements of any of them or an alleged invention. Based on the Ayyangar report, certain changes were made to the definition of patentable inventions. Sections 3, 4 and 5 list the inventions that are not currently patentable in India. They are reproduced below.

> The following are not patentable inventions within the meaning of this Act:
> **Section 3**
> (a) an invention which is frivolous or which claims anything obviously contrary to well established natural laws;
> (b) an invention, the primary or intended use of which would be contrary to law or morality or injurious to public health;
> (c) the mere discovery of a scientific principle or the formulation of an abstract theory;
> (d) the mere discovery of any new property or new use for a known substance or of the mere use of a known process, machine or apparatus unless such known process results in a new product or employs at least the new reactant;
> (e) a substance obtained by a mere admixture resulting only in the aggregation of the properties of the components thereof or a process for producing such substance;
> (f) the mere arrangement or re-arrangement or duplication of known devices each functioning independently of one another in a known way;
> (g) a method or process of testing applicable during the process of manufacture for rendering the machine, apparatus or other equipment more efficient or for the improvement or restoration of the existing machine, apparatus or other equipment or for the improvement or control of manufacture;
> (h) a method of agriculture or horticulture;
> (i) any process for the medicinal, surgical, curative, prophylactic or other treatment of human beings or any process for a similar treatment of animals or plants to render them free of disease or to increase their economic value or that of their products.

Section 4

Inventions relating to atomic energy are not patentable – no patent shall be granted in respect of an invention relating to atomic energy falling within sub-section (1) of Section 20 of the Atomic Energy Act, 1962 (33 of 1962).

Section 5

Inventions where only methods or processes of manufacture are patentable: In the case of inventions:

(a) claiming substances intended for use, or capable of being used, as food or as medicine or drug, or

(b) relating to substances prepared or produced by chemical processes (including alloys, optical glass, semi-conductors and inter-metallic compounds),

Patent shall be granted in respect of claims for the substances themselves; aims for the methods or processes of manufacture shall be patentable.

It is also pertinent to list section 2(1) which defines the terms 'medicine' or 'drug' from the point of view of patentability;

(i) all medicines for internal or external use of human beings or animals;

(ii) all substances intended to be used for or in the diagnosis, treatment, mitigation or prevention of diseases in human beings or animals;

(iii) all substances intended to be used for or in the maintenance of public health, or the prevention or control of any epidemic disease among human beings or animals;

(iv) insecticides, germicides, fungicides, herbicides and all other substances used for or intended to be used for the protection or preservation of plants;

(v) all chemical substances which are ordinarily used as intermediaries in the preparation or manufacture or any of the medicines or substances above referred to

The definition of 'drug' under section 2(1)(i) is not exhaustive. The term 'drug' is also defined under section 2(b) of the Drugs and Cosmetics Act 1940 as including:

(i) all medicines for internal or external use of human beings or animals and all substances intended to be used for or in the diagnosis, treatment, mitigation or prevention of disease in human beings or animals and

(ii) such substance (other than foods) intended to affect the structure or any function of the human body or intended to be used for the destruction of vermins or insects which cause disease in human beings or animals, as may be specified from time to time by the Central Government by notification in the official gazette.

The Central Government has, by notification 56a (Government of India, 1961), specified that contraceptives and the following disinfectants should also be classed as drugs:

1. Disinfectant fluids made from coal-tar oils, coal-tar acids or similar acids derived from petroleum with or without hydrocarbons.
2. Disinfectant fluids made from synthetic or naturally occurring substances other than those mentioned in paragraph (1) above by virtue of their composition possessing disinfectant properties or with claim to possess disinfectant properties.

The 'food and agriculture' sector is an interdisciplinary area with links to other fields such as chemical industry, environment, processing technologies, water sources and its management, energy utilization, storage facilities, distribution, transportation, etc. Innovations in any of these fields can contribute to improvements in productivity, efficiency and cost-effectiveness in the agricultural sector.

The restrictions listed above immediately affect the protection of inventions in chemical industries related to agrochemicals, drugs, pesticides, insecticides, agricultural practices, plant breeding, seed development, highly specialized materials, etc., including any products resulting from chemical reactions. In these cases only processes are patentable. Similarly, section 3(e) defines the type of formulations which fall outside the scope of patentable inventions. Any product formulation in which the ingredients do not exhibit synergy in some of the claimed properties cannot be patented in India. Therefore inventors have to design experiments to clearly illustrate synergic properties. However novel a formulation is, if the properties of the ingredients are merely additive then this is not adequate to satisfy section 3(e). Similarly, implements such as a plough/thresher can be patented so long as they are not interpreted as falling in the class of 'method of agriculture'. Thus such specifications must be drafted with extreme care and complete awareness of such special features in the Act.

Apart from these restrictions there are a few special provisions in the Indian Patent Act 1970 defining the conditions under which Patent can be considered to have been worked in India and giving guidelines for the issue of compulsory licenses, licenses of right and revocation especially with respect to the powers vested with the government of India. These are discussed below.

Other unique features of the Act

PROCEDURAL ASPECTS. Residents in India first have to apply for a patent in India and must then wait for six weeks before filing a corresponding patent application in any foreign country. If one wishes to proceed with the foreign filing before the expiry of this six week period, an application must be made to the controller for a waiver. Only on receipt of the waiver can this foreign filing be done before the six week waiting period.

The applicant gives an undertaking that, up to the date of the acceptance of the Indian Patent application, they would keep the controller informed in writing from time to time on the status of any corresponding foreign filings and, if required by the controller, furnish the details of the objection, if any, taken on such foreign applications on the grounds that the invention is lacking in novelty or patentability, amendments affected in the specification and the claims allowed in respect thereof. Failure to do so can be grounds for opposition and revocation of the patent.

The provisional specification must describe the invention sufficiently, preferably with an adequate number of examples. No claims need be written in the provisional specification. The complete specification, which must be filed within 12 months (or a maximum of 15 months if permission has been given by the controller for late filing), must include the claims which are within the scope of the invention described in the provisional specification.

The claim or claims of the complete specification must relate to a single invention. The controller can direct the applicant to make a divisional application if he/she considers that the claim or claims in the complete specification relate to more than one invention.

The complete specification is published in the *Gazette of India*, Part III, Section 2, only after the application has gone beyond the acceptance stage. In India there is no publication if 18 months have elapsed since the provisional specification was filed. When a provisional or complete specification is filed the *Gazette* publication carries only the title, inventors, assignees and the date of filing.

The term of a patent is 14 years from the date of submission of the complete specification. However, for inventions claiming the method or processes of manufacture of a substance which is intended for use as, or is capable of being used as, food/medicine or drug (as defined earlier) the term of the patent is five years from the date of sealing of the patent or seven years from the date of submission of the complete specification, whichever period is shorter.

WORKING OF PATENTS, COMPULSORY LICENSING AND REVOCATION. Grant of a patent is subject to the following special conditions:

1. The patented product may be imported by or on behalf of the government for its own use.
2. Any patented process may be used by or on behalf of the government.
3. In the case of a patent with respect to a medicine or drug the government can import the medicine or drug for the purpose of its own use or distribution in the public health system; such use is to be notified in the *Gazette*.

In the public interest, the government can direct the revocation of a patent if it considers the patent to be damaging to the state or generally prejudicial to the public. In such cases the patentee is given a hearing prior to issuing an order and this is announced in the *Gazette*.

Importation of any article patented into India (this also includes a process patent) does not amount to working of the patent in India. When a patented invention has been used commercially in India by the patentee he is said to have 'worked the patent'. For example, a machine patented in India has to be manufactured in India in order to be considered as having worked the patent. Similarly, a process for manufacturing a drug may have been patented in India, but importing the drug does not amount to working of the patent. This ruling was introduced to encourage growth of industries in India.

Once three years have expired since the date of sealing of a patent, any person may apply to the controller for the grant of a compulsory license on the grounds that the reasonable requirements of the public have not been satisfied or that the invention is not available to the public at a reasonable price. This provision can be evoked if the patentee is

- not exploiting the invention to the fullest extent possible;
- unable to manufacture the article for adequate supply on reasonable terms;
- refusing to grant licenses on reasonable terms thereby obstructing existing trade, commercial activities or industry, and their development in India;
- meeting demand in India through importation of the patented article;
- not initiating infringement action on 'persons' importing the patented article in India;
- not meeting export demand of the patented article by manufacturing in India.

Records show that this provision has seldom been exploited in India since the Patent Act 1970 came into force.

Endorsement of a patent with the words 'licenses of right' may be ordered by the controller based on an application made by the Government at the end of three years from the date of sealing of the patent on the grounds that reasonable requirements of the public have not been met (as defined above) or that the invention is not available to the public at a reasonable price. All patents that fall under section 5 of the Patents Act 1970 are automatically endorsed with the words 'licenses of right' three years after the date of sealing of the patent. As discussed above, these include process patents in the areas of drugs, foods and agrochemicals. The effect of this provision is that the patentee is required to grant a license to anyone who wants to work the patented invention in India. If the parties are unable to agree reasonable terms, they may apply to the controller to settle the issue. The royalty must in no case exceed four percent of the net ex-factory price in bulk of the patented article (exclusive of taxes levied under any law for the time being in force and any commission being payable).

The controller has the power to revoke patents for non-working in India. For example, where, in respect of patents, a compulsory license has been granted or has been endorsed with the words 'licenses of right', the controller can, on receiving an application (at the end of two years from the grant of a compulsory license or the grant of the first license under conditions of license of right) from the government or from any interested person(s), revoke the patent on the grounds that reasonable requirements of the public have not been met or that the invention is not available to the public at a reasonable cost. These provisions were introduced into the Indian Patents Act 1970 to protect national interests, to help promote diverse industries in India and to control monopolistic ventures.

In the area of enforcement, the burden of proof (to prove infringement of a patent) is on the patentee. This feature has been debated extensively and is currently under review.

Patenting in India – a profile

A quick scan of the data on patents filed, granted and in force (Figs 10.3 and 10.4) drives home the point that, in India, patenting has not been a

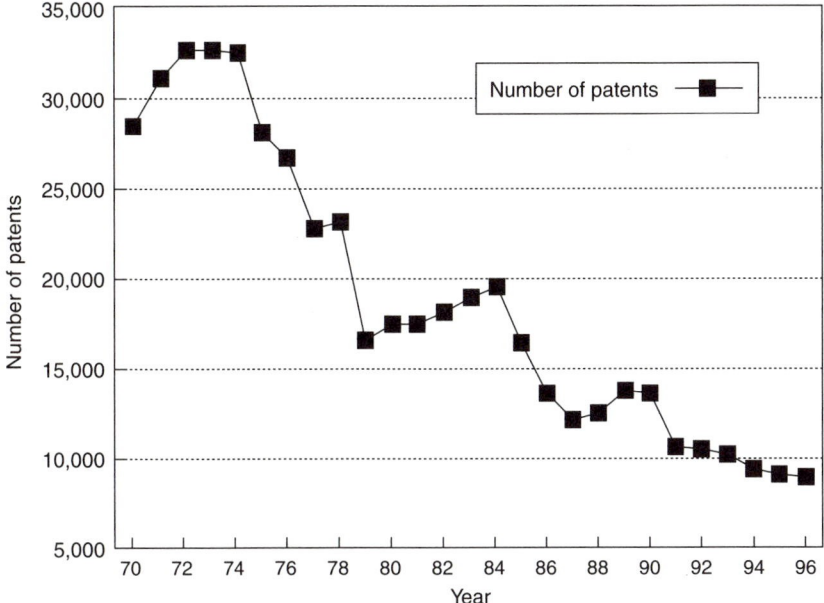

Fig. 10.3. Number of patents in force in India in the period 1970–1996.

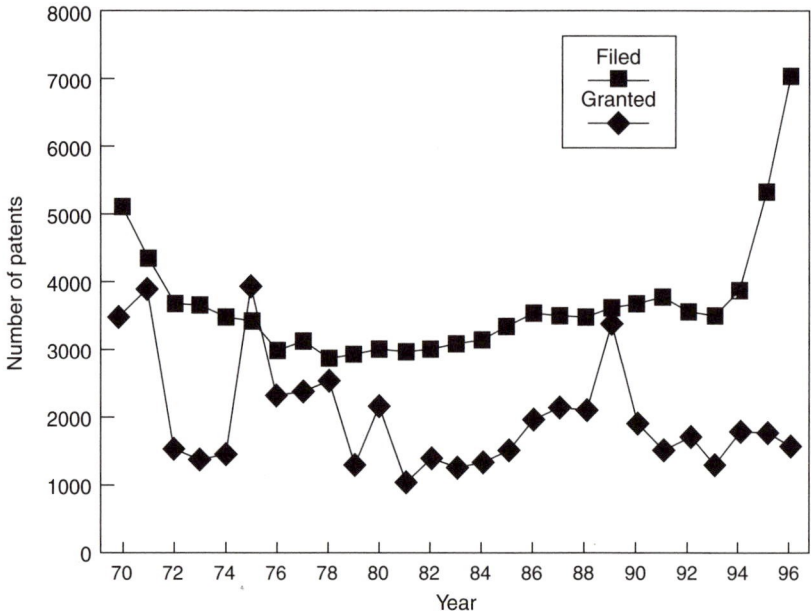

Fig. 10.4. Number of patents filed (■) and number granted (♦) in India in the period 1970–1996.

major activity in the last 25 years, in contrast with most industrially developed countries. It is significant that significantly fewer patents were filed and granted after 1972 than in 1970 and 1971. This may be considered to be a direct impact of the Patents Act 1970 which came into effect in April 1972. The number of patents in force has also gradually fallen over the years.

Tables 10.1 to 10.3 show the numbers of patents applied for, granted and in use, classified according to field. Most patents were filed in the 'chemicals' category, and the fewest in the drugs/food-related area. On average, 42% of patent applications in the chemicals category are granted. Most of these are inventions in the field of processes related to organic chemicals and polymers. Patents in methods of manufacture of drugs, drug intermediates or pharmaceutical formulations have also been popular. Slow-release or controlled release formulations are gradually gaining significance. Among other requirements such formulations must necessarily satisfy the conditions laid down by sections 3(d) and (e) of the Act.

These trends can be understood in terms of the restrictions contained in the Patents Act 1970 and the attitude of the Indian industries, who considered it more appropriate to enter collaborative arrangements or import ready-made technology for quick exploitation in a fairly secure national market. Industrial R&D centers in India have gradually become skilled at working around patents, especially those relating to

Table 10.1. Patent applications filed from 1986 to 1996 in various fields.

Year	Chemical	Drug	Food	Electrical	Mechanical	General	Total
1986–7	1112	214	34	577	972	579	3489
1987–8	1020	198	39	563	847	772	3598
1988–9	1191	221	21	419	974	772	3598
1989–90	1225	216	13	454	932	821	3661
1990–1	1297	258	41	492	1173	502	3763
1991–2	1185	323	38	468	994	544	3552
1992–3	1138	234	29	461	946	659	3467
1993–4	1122	273	82	426	895	1071	3869
1994–5	1516	692	125	653	1337	1030	5330
1995–6	1934	1000	104	1131	1599	540	7036

Table 10.2. Patents granted from 1986 to 1996 in various fields.

Year	Chemical	Drug	Food	Electrical	Mechanical	General	Total
1986–7	474	185	21	293	705	448	2125
1987–8	731	124	15	228	489	517	2104
1988–9	704	300	44	608	1183	541	3380
1989–90	389	300	44	322	692	321	1890
1990–1	339	87	10	285	535	235	1491
1991–2	474	118	10	167	181	726	1676
1992–3	318	94	12	194	372	282	1272
1993–4	436	145	30	132	215	788	1746
1994–5	642	232	49	135	161	540	1759
1995–6	470	132	34	56	159	682	1533

From: *Patents* – Annual Report of the Controller General of PDTM under section 155 of Indian Patents Act, 1970 (1992–1996).

Table 10.3. Number of patents worked from 1987 to 1991.

Year	Chemical	Mechanical and general	Electrical	Total
1987	151	318	89	558
1988	155	301	118	574
1989	64	172	53	289
1990	80	134	38	282
1991	93	210	40	343

From: *Patents* – Annual Report of the Controller General of PDTM under section 155 of Indian Patents Act, 1970 (1992–1993).

chemical processes, and most foreign companies who had earlier filed patents in India no longer find it worthwhile continuing to file patents or even to renew their patents here.

There are positive features too, though. Some of the National Laboratories over the years have built up a strong expertise base in chemical technology and product/processes have been developed, patented and transferred to Indian industries especially in the fields of organic chemicals, inorganic catalysts, etc. Transfer of technologies from these national R&D centers to companies in other countries is a growing trend in the last few years.

THE WAY AHEAD

Policy changes and foreign direct investment

With the declaration in 1991 of intent to open up the Indian economy, significant changes have been initiated in Indian economic and industrial policies. Two changes that have had immediate impact on the agriculture, foods, drugs/pharmaceutical sectors are as follows:

1. Hybrid high-yielding seeds, tissue culture propagation of elite plant materials, biofertilizers and biopesticides are classified as industrial activities.

2. Bulk drugs involving use of recombinant DNA technology, and bulk drugs requiring *in vivo* use of nucleic acids as active principles and formulation based on use of specific cell- or tissue-targeted formulations, would require compulsory licensing.

These have already set the scene for development of biotechnology-based industries in India. Extensive national debates on various issues related to IPR and the Indian Patents Act 1970 have also taken place in the last few years. These have helped to enhance the level of patent literacy considerably. The first impact has been on the amount and type of foreign direct investments that have flowed into India since 1991. A snapshot is presented in Figure 10.5. Foreign direct investments are still low compared with those in most neighboring Asian countries such as China, Singapore and Korea. The potential for investment is large and it is expected to increase sharply with further liberalization of industrial policies and harmonization of the IPR laws in India.

Towards a more harmonized IPR system

India is a signatory to GATT and is therefore expected to implement the clauses under TRIPs to harmonize the Indian IPR laws with the rest of the global partners in the World Trade Organization (WTO). India's copyright laws do not need amendment as they are already harmonized with international ones. The Trademarks and Merchandise Act 1958,

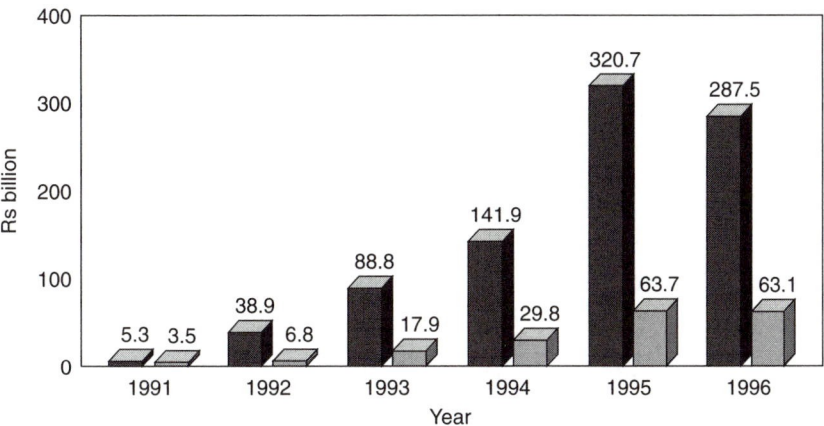

Fig. 10.5. Foreign direct investments (FDI) in India over the period 1991–1996. ■, FDI approvals; ▨, actual inflow. From the *Times* of India; 22 January 1997.

the Indian Designs Act 1911 and the Patents Act 1970 are due for revision, though, and this is currently under consideration. India is a member of the WTO but has not yet joined the Paris Convention. It is therefore not a party to the Patent Cooperation Treaty (PCT) and consequently also not a member of the Budapest Treaty.

Possible changes in the patent laws

Any changes would have to address the unique features of the Indian Patents Act 1970, which were discussed above. A proper balance between national/public interests on one hand and private interests on the other will have to be struck before a bill is considered to be debated in the parliament.

The Patent Ordinance 1994 no. 13. of 1994 (Government of India, 1994; Ganguli, 1995) as supplementing the Indian Patents Act 1970 came into force on 31 December 1994 for a short period but was rejected by the upper house of the Indian Parliament in March 1995. This ordinance, which is no longer in force, made provisions for filing of product patents for foods, drugs and medicines, including insecticides, pesticides, herbicides, etc. In fact, it virtually removed all restrictions in section 5(a). Thus every item listed in section 2(1) 1 was included under the banner of patentable inventions.

The ordinance broadened the 'convention countries' for the purpose of patenting to include all the member countries of the WTO. This meant that the priority date of a patent application filed in any of the member countries could be used to file a patent in India within 12 months of its filing in those countries. The ordinance also made provisions for the required protection of such products under an arrangement of exclusive marketing rights for a period of five years for those products which are given product patents in any other member countries and are introduced in the market, provided an application is made to that effect to the Government of India in the prescribed manner. It further considered importation of a product into India as having worked a patent granted in India even if the product was not manufactured in India. However, the ordinance gave powers to the Government to intervene while granting exclusive marketing rights or issuing compulsory licenses. It also waived the clause that residents in India must file their inventions in India before filing them abroad.

Since the rejection of this patent amendment bill, the matter is with the special committee in Parliament. The committee has been deliberating on the integrated issues of IPR, which include amendments to the Patents Act, the Merchandise and Trade Marks Act, a *sui generis* system and a Plant Varieties Protection Act. It is expected that bills recommending these changes will soon be presented to Parliament. A few of the possible changes are listed in Table 10.4.

In the process of harmonization of the patent laws, one will have to consider the options provided by Article 27 of TRIPs, which states:

27(2) Members may exclude from patentability inventions, the prevention of the commercial exploitation of which within their territory is necessary to protect the public or morality, including to protect human, animal or

Tabl 10.4. Possible modifications to the current Patent Act.

Present Act	Possible modifications
Product patent not allowed for pharmaceuticals, food products or agrochemicals, only for process patents	Product patents to be allowed in all fields of technology
Duration 14 years for all areas except in pharmaceuticals, food products and agrochemicals where it is 7 years	Uniform duration of 20 years
Government has powers to grant compulsory licenses	Compulsory licenses to be given only on merits of each case. Patent holder will be given opportunity to be heard
Importation does not amount to working of patent	No discrimination between imported and domestic products
For process patents, in case of infringement, burden of proof on plaintiff	Burden of proof on alleged infringer

plant life or health or to avoid serious prejudice to the environment, provided that such exclusion is not made merely because the exploitation is prohibited by law.

27(3) Members may also exclude from patentability:

(a) Diagnostic, therapeutic and surgical methods for the treatment of human or animals;

(b) Plants and animals, other than micro-organisms and essentially biological processes for the production of plants or animals other than non-biological and microbiological processes. However, members shall provide for the protection of plant varieties either by patents or by an effective *sui generis* system or by any combination thereof. The provisions of this sub-paragraph shall be reviewed four years after the date of entry into force of the WTO agreement.

Thus in the agriculture sector, there is no obligation to patent any material, but there is a requirement to provide protection to new varieties, which is the trend in many countries.

The implications of the new IPR regime will have to be examined as soon as the laws are passed and come into force. However, if the trends in the last few years serve as a broad indicator, it appears that India is setting up a platform for a harmonized IPR system for promotion of industrial and innovative activity.

ACKNOWLEDGEMENT

The author is grateful to the editor of Universities Press (India) Ltd, Hyderabad, India for granting permission to reproduce Figures 10.1, 10.2 and 10.5 from Ganguli (1998).

REFERENCES

Ayyangar, R. (1959) *A Report on the Revision of Indian Patent Law.*

Bose, D.M., Sen, S.N. and Subbarayappa, B.V. (eds) (1971) *A Concise History of Science.* Indian National Science Academy, New Delhi.

Chand, B.T. (1950) *Report of the Patents Enquiry Comittee (1948–1950).*

Ganguli, P. (1995) Commentary on Patent Ordinance 1994, No. 13 of 1994. Supplementing the Indian Patent Act 1970. *World Patent Information* 17(3), 193–194.

Ganguli, P. (1998) *Gearing up for Patents – the Indian Scenario.* Universities Press (India) Ltd, Hyderabad.

Government of India (1961) Ministry of Health Notification No. 1-20/60-D dt. (6 March 1961).

Government of India (1994) Patent Ordinance 1994, No. 13 of 1994. Supplementing the Indian Patents Act 1970. In: *Gazette of India*, Part II, Section 1, number 81.

Narayan, P. (1985) *Patent Law*, 2nd edn. Eastern Law House Pvt Ltd , Calcutta.

APPENDIX 10.1. THE INDIAN DESIGNS ACT 1911

The Designs Act of 1911 (amended on 1 April 1978) allows registration of features of shape, configuration, pattern or ornamentation applied by any industrial processes or means whether manual, mechanical or chemical, separate or combined which in the final article appeal to and are judged solely by the eye. This does not include any mode or principle of construction or anything which is in substance a mere mechanical device and does not include any trademark as defined in the Trademark and Merchandise Marks Act of 1950.

For the purpose of registration of designs, goods to which the designs are to be applied are divided into 14 classes, as listed in the table below.

If a design could potentially be included in more than one of these classes, it is always recommended that it be registered in all those classes. The registration is valid for five years from the date of application. It is possible to extend the term by two further periods of five years on application before the expiry of the copyright of the design. In the case of a convention country application where the priority is allowed, the date is five years from the priority date.

The design should be new or original and previously published in India at the date of application for registration.

Applications for registration of designs in India are to be made to

Class	Description
1	Articles composed wholly of metal or in which metal predominates, and jewellery
2	Books and book binding of all materials
3	Articles composed wholly of india-rubber, wood, bone, ivory, papier mâché, celluloid, Bakelite, or like substances, or of materials in which such substances predominate (except articles included in Class 10)
4	Articles composed wholly of glass, earthenware or porcelain, clay (burnt or baked) or cement or in which such materials predominate
5	Articles composed wholly of paper, cardboard, millboard or strawboard (except articles included in Class 2 and paper hangings), or in which such materials predominate
6	Articles composed wholly of leather or in which leather predominates, not included in other classes
7	Paper hangings
8	Carpets, rugs and floor coverings in all materials
9	Lace
10	Boots, shoes and the like footwear
11	Millinery and wearing apparel (except articles included in Class 10)
12	Goods not included in other classes
13	Printed or woven designs on textile goods other than checks or stripes
14	Printed or woven designs on textile goods being checks or stripes

the Controller of Patents and Designs at Nizam Palace, New MSO Building (sixth floor), The Patent Office, 234/4, Acharya Jagadish Bose Road, Calcutta 700 020, India.

Clear statements on the novelty of the designs in terms of its shape and/or configuration/ornamental surface/pattern must be indicated. The novelty must be clearly marked in the design submitted. The application must be accompanied by an appropriate disclaimer so that these items are not confused with any trademark or any previously registered design. Other disclaimers can be used, for example: 'No claim is made by virtue of this registration in respect of any mechanical or other action of the mechanism whatever or in respect of any mode or principle of construction of the article.'

If the representation contains words, letters, numerals, etc. a disclaimer should be inserted in the following manner: 'No claim is made by virtue of this registration to any right to the exclusive use of the words, letters, numerals, flags, crowns, arms, etc. appearing in the design.'

If the representation contains different colors but these are not features of the design a disclaimer should be inserted in the following manner: 'No claim is made by virtue of the registration to any right to the exclusive use of color or color combination as appearing in the design. Extraneous matter not forming part of the design shall be removed from the representation.'

The design application is examined by the Controller and objections, if any, are raised with the applicant, who is expected to make the necessary changes to satisfy the Controller. Once the Controller is satisfied, the design is accepted and notified in the *Gazette of India*, Part III, Section 2, and a certificate is issued on the registered design. A registered design is not open to public inspection for a period of two years from the date of registration except by its proprietor or a person authorized by him or a person authorized by a controller or by a court or by a person who has been refused registration of a design on the ground of identity with the design already registered. All the procedures for environment of patents are followed for enforcement of registered designs.

APPENDIX 10.2. SELECTED CENTERS FOR PATENT INFORMATION

National Informatics Centre, Patent and Knowhow Information Division, A Block, CGO Complex, Lodhi Road, New Delhi 110 003.

Patent Management Division, CSIR, 14 Satsang Vihar Marg, Special Institutional Area, New Delhi 110 067.

NICHEM/NCL, National Chemical Laboratory, Pune 411 008, Telex: 0415 266; E-mail: kri@ncl.ernet.in, nichem@ncl.ernet.in, ncl@iucaa.ernet.in

National Research and Development Council (NRDC), 20–22 Zamroodpur, Kailash Colony, New Delhi 110 048
Patent Information System (PIS), CGO Complex, 'C' Block, Seminary Hills, Nagpur 440 006.
IPR CELL IN Department of Electronics, R&D Policy & Coordination Division, Department of Electronics, 6, CGO Complex, New Delhi 110 003.

APPENDIX 10.3. ADDRESSES OF PATENT OFFICES IN INDIA

Head Office (design applications are filed only in this office)

Nizam Palace, 2nd MSO Building, 5th, 6th and 7th Floors, 234/4, Acharya Jagdish Bose Road, Calcutta 700 020. Fax: 91-33-2473851; Tel.: 91-33-2474401.

Branch Offices

Todi Estates, Third Floor, Lower Parel (West), Bombay 400 013. Fax: 91-22-4925094; Tel.: 91-22-4924058 or 4925092.
Units 401–405, Third Floor, Municipal Market Building, Saraswathi Marg, Karol Bagh, New Delhi 110 005. Fax: 91-11-5716204; Tel.: 91-11-5716209.
61 Wallajah Road, Madras 600 002. Fax: 91-44-841014; Tel.: 91-44-845324.

Other Offices for Patent-related Activities in India

Patent Information System, Third Floor, Block 'C', CGO Complex, Seminary Hills, Nagpur 400 006. Fax: 91-712-528186; Tel.: 91-712-525670.
Office of the Controller-General of Patents, Designs & Trademarks, 101, M.K. Road, Bombay 400 020. Fax: 91-22-2053372; Tel.: 91-22-2039050.

APPENDIX 10.4. PATENT FACILITATING ORGANIZATIONS IN INDIA

(This is not an exhaustive list but includes those organizations which have been active recently.)

National Research & Development Corporation

20 Zomroodpur Community Centre, Kailash Colony Extn, New Delhi 110048. Tel.: 91-11–641–8615/7821; Fax: 91-11-644940/6460506; E-mail: nrdc@x400.nicgw.nic.in
Promotes commercialization of Indian inventions, including financial and technical assistance to file and execute patents in foreign countries. Also offers patent search facilities.

Technology Information, Forecasting and Assessment Council (TIFAC)

Department of Science and Technology, Technology Bhavan, New Mehrauli
Road, New Delhi 110016. Tel.: 91-11-667373/405, Fax: 91-11-6863866
Patent-facilitating organization to create awareness, IPR and help scientists
to patent their inventions by providing financial and technical support.
Also provides patent information as inputs into various R&D programs.

Department of Scientific and Industrial Research (DSIR)

Technology Bhavan, New Mehrauli Road, New Delhi 011016. Tel.: 91-11-
662626/667373; Fax: 91-11-655145, 6862416.
The DSIR promotes technology evaluation and filing of patents of Indian
inventions. Under the National Information System for Science and
Technology (NISSAT) it has set up a number of information centers aimed
at information, documentation, patent awareness, patent searches, etc.

Intellectual Property Management Division

INSDOC Building, 14 Satsang Vihar Marg, New Delhi 110067. Tel.: 91-11-6962560,
91-11-6968819; Fax: 91-11-6968819; E-mail: root@ocsptntu.ren.nic.in
Under the Council of Scientific and Industrial Research, this organization
provides support to all the 40 National Laboratories in India on IPR issues,
which include patenting of inventions from these laboratories, enhancing
awareness on IPR, training scientists to identify patentable inventions,
patent information, etc.

11

Russia

Andrei A. Baev

Mayer, Brown and Platt, 25th Floor, 350 South Grand Avenue, Los Angeles, CA 90071-1503, USA

INTRODUCTION

After the collapse of communism in the former Soviet republics and the end of the Cold War, the emerging Russian market has attracted much interest and attention from foreign investors. When a foreign company considers investing in Russia, however, there are many factors it must consider. Protection of intellectual property rights (IPR) is one of these.

There are four general situations in which foreign companies encounter problems of IPR protection under Russian legislation: (i) when a foreign company brings its own intellectual property (IP) to Russia; (ii) when joint IP is developed during the activities of a cooperative venture; (iii) when a Russian partner brings IP to the joint venture in Russia; and (iv) when a foreign company utilizes Russian IP exclusively outside of Russia. While the first two situations relate to the enforceability of existing Russian laws, the latter two mainly reflect the drawbacks of the former Soviet legislation as applied to the IP predating the current legislation.

Russia has a long and widespread tradition of disregarding IPR. The deficiencies in protecting IPR in Russia are partly rooted in the absence of a tradition of private ownership in the former Soviet Union. The Soviet laws stemmed from the uniquely communist idea that societal benefit in the form of cultural and scientific development took priority over an individual's rights. Accordingly, the primary goal of IP law was to disseminate creative works and inventions to a wider sector of the population rather than to curtail undue competition and protect the individual rights of creators.

© CAB INTERNATIONAL 1998. *Intellectual Property Rights in Agricultural Biotechnology* (eds F.H. Erbisch and K.M. Maredia)

One of the goals of legal reform in post-Soviet Russia has been to bring Russian IP law into accord with Western practice. However, although most of the deficiencies of the Soviet-type system of IPR have been eliminated and although the new legislation in Russia does address domestic IP protection on a par with international standards, two major problems of protecting IPR in Russia remain: (i) clarification of the IPR of various legal entities, collectives and individual inventors with respect to intellectual and industrial property predating the current legislation; and (ii) enforcement of the existing legislation. Although, with the passage of time, the problems associated with IP developed prior to the new laws will become less critical in some, though not all, cases. The problem of enforcing the laws has become the stumbling block for the implementation of the new standards of IP protection in Russia. The need for improvement in Russia's enforcement of IPR is the greatest obstacle for Russia in its bid for admission to the World Trade Organization (WTO). Neither the Russian people nor government institutions yet realize the need to protect IPR. The Russian government attaches very low profile to the protection of such rights since it considers economic and social stabilization to be a priority.

The purpose of this chapter is to examine some of the recent changes taking place today in Russian intellectual and industrial property law and to explore existing obstacles to implementing and enforcing the law. The chapter discusses three general areas of IP law related to the general theme of this handbook: patents, protection of plants and animals and trademarks.

THE PATENT LAW OF THE RUSSIAN FEDERATION

Generally, patent protection promotes new discoveries by granting a limited monopoly on use of inventions before they enter the public domain. Besides promoting the general progress of science and useful arts, patent law serves three major functions: (i) stimulation of research by providing researchers with a monopoly on their innovations; (ii) fostering efficient allocation of resources so as to prevent duplication of research; and (iii) identification of the legal owner of particular inventions. Since Soviet patent law failed to achieve these objectives, rectification of this situation was at the heart of the patent law reform in the post-Soviet Russia.

Protection of intellectual property predating the current legislation

In contrast with the Western legal model, the Soviet system for inventions was characterized by collective ownership of certified inventions protected by inventors' certificates of authorship (*avtorskoe svide-tel'stvo*). An inventor's certificate offered a guarantee of a fixed royalty-like payment and certain other material and immaterial rights and privileges for the inventor, whose authorship was officially certified, and it entitled the Soviet state to organize freely the commercial exploitation of the invention. Inventions protected by inventors' certificates could be liberally exploited without any permission by the Soviet state, cooperatives, public enterprises and other legal entities. The inventions for which an inventor's certificate were issued were thus placed in a public domain. The state also monopolized control over the use of Soviet inventions outside the USSR by entering into licensing agreements with regard to the exploitation of inventions abroad and by fulfilling various preparatory tasks for such exploitation, such as patent applications and patent litigation.

However, an inventor's certificate was not the only legal technique employed to protect the intellectual rights of Soviet inventors. Inventors were granted a traditional 'freedom of choice' to protect their inventions either by an inventor's certificate or by a patent. In contrast with the 'public domain' status of the inventions protected by an inventor's certificate, a patent provided patentees with the exclusive rights of utilizing their inventions. Although Soviet law provided for patent protection, in practice only foreigners opted for patents. Patents were extremely costly to apply for, maintain, market and defend, in addition to not being freely exploitable. Indeed, in a socialist economy the incentive structure was such as to make individual inventors favor an inventor's certificate over a patent as the means of protecting their rights. Besides the guaranteed payments for the exploitation of their inventions, the inventors also avoided the expenses related to the vending of their inventions, and they were provided with substantial social benefits as well, such as the right to be named as the author and advantages of employment, promotion, preferential admission to universities and graduate schools, additional housing space, etc., none of which were extended to the holders of patents. Furthermore, while there were filing and issuance fees for patents, inventors' certificates did not require such fees. If a patent application was rejected, the fees and expenses were imposed on the applicant, but there were no fees for an application to issue an inventor's certificate. The failure to pay annuities terminated patent protection as well. A patent could be opposed and cancelled during the entire period of protection. In contrast, an inventor's certificate could be contested only in the first year after its issuance. Thus, the dominant means of protecting the IPR of individual inventors in the former Soviet Union was an inventor's certificate.

Patents were not available in the most important practical case of employee inventions (*sluzhebnye izobreteniia*), for which only an inventor's certificate could be issued. According to some data, at least

80% of all Soviet inventions were employee inventions. The state was considered the holder of all IPR to the scientific/technical, industrial or other products of the state research and development (R&D) institutes and state enterprises, as long as the product was created by the engineers and technical staff of these enterprises during their employment, in connection with the work of the inventor in a state enterprise, on the enterprise's facilities, upon the direction of such enterprises, or by utilizing government funds, which had been apportioned to the enterprise by the state. Under these circumstances, the overwhelming majority of the inventions and know-how developed in the Soviet Union were classified as 'employee inventions', for which IPR belonged to the state.

The overlap of the old Soviet legislation with the new Russian law on IP presents a number of problems. The inventors' certificates predating the current legislation still remain in effect if they have not been exchanged for patents. Thus, the use of the inventions protected by inventors' certificates is not exclusive to the holders of the certificates and in practice these inventions can be freely utilized by the state or any private company without compensation. Foreign businesspersons should therefore insist that a Russian partner exchange its inventors' certificates for patents under the new Russian law. In addition, foreign companies should always enquire about a Russian company's outstanding obligations to individual inventors. If such obligations exist, the company should insist that they be fully satisfied by the Russian partner.

Moreover, as a result of the recent privatization of state enterprises and R&D institutes, it is currently unclear who holds patent rights to the inventions that formerly belonged to the state-owned enterprises. Reorganization of the State Industrial Ministries, abolition of the industrial/technical consortia, and restructuring of privatized state enterprises has only complicated the problem of recognizing patent holders. In addition, a great number of the modern technological processes and inventions currently used in privatized companies were developed at a time when they were only protected by inventors' certificates. While patents are usually transferred with other assets during privatization, inventors' certificates cannot be conveyed to a privatized enterprise since the state enterprise never had exclusive rights to the inventions protected by inventors' certificates in the first place. In practice, any private company, public organization or state entity may assert a claim of IPR to the inventions protected by such certificates, because the inventions are presumably owned by all the public. This becomes particularly important if a Russian company is planning to export a technology that promises to provide a handsome income. The government may also claim its property right to the IP in order to protect national strategic interests, obstruct suspicious projects or bargain with large foreign investors.

According to civil law, a privatized enterprise assumes all obligations of the state enterprise being privatized from the time the privatization is completed. However, due to the public domain status of most Soviet inventions, a privatized enterprise may not be able to assert rights to its inventions if the privatization of the enterprise was not conducted properly. An infringer of the IPR of such an enterprise can always contest the claim of infringement on the basis of the IPR never having passed to the enterprise because of the violation of certain privatization formalities. Therefore, it is sometimes more important to analyze the process of privatization rather than the corporate papers of a new privatized enterprise.

In situations where a Russian partner of a joint venture is a former state-owned entity that has been recently privatized, it is highly advisable to comply with the following strategy:

1. Enquire into the Decision of State Property Committee (*Goskomimushchestvo*) regarding privatization of the enterprise.
2. Verify the fulfillment of the legal formalities reflected in the privatization plan.
3. Check the contract of purchase regarding the IPR provisions if the enterprise was privatized by way of leasing the enterprise with the right of purchase or by selling the enterprise at auction or by comparative bidding (*konkurs*).
4. Inspect the internal records (*balance*) of the enterprise to determine whether patents and other IPR are recorded as part of the enterprise's assets.

However, IPR are only very rarely clarified in either the charter of the privatized enterprise, or its privatization plan. The best course of action in such a situation would be to insist that the Russian partner disclose in the joint-venture agreement all its legal rights and obligations with regard to the IP in its possession, including the representations and warranties: (i) that it has a right of conveyance; (ii) that no outstanding encumbrances affect this right; and (iii) that the Russian partner will remedy all possible legal defects and disputes that might arise in the future with regard to the provenance of these rights. The Russian partner should account for its existing obligations to the inventors, and disclose whether it possesses exclusive use of all inventions. To further protect one's interests in the transferred rights to an invention, one should insist on a special provision imposing liquidated damages on a Russian partner in a case of possible cancellation of these rights in the future.

The new Patent Law of the Russian Federation

The Patent Law of the Russian Federation, which places the protection of IP in Russia on a par with international standards, was adopted on 23 September 1992, and became effective on 14 October 1992. A short overview of the Patent Law of the Russian Federation is given in the following sections.

Subject matter of the law
The Patent Law covers inventions, utility models and industrial designs.

PATENTS. Like US patent law, the Russian law limits patentable inventions to those that possess three characteristics: novelty, inventiveness and industrial applicability. An invention is defined as novel if it is not known from prior art, including any kind of information published anywhere in the world and made available to the public, as well as information available from prior Soviet patents, inventors' certificates, or prior applications for patents or certificates. 'Inventiveness' is established if, for one skilled in the art, the invention does not obviously proceed from prior art. As some commentators note, since there are no qualifying words, this standard may become somewhat more subjective than that in US patent law, although the Russian Patent Office regulations are expected to contain instructions which should clarify the non-obviousness criterion. 'Industrial applicability' refers to the invention's practical usefulness in industry, agriculture, public health services and other fields of activity. Besides the inventions analogously patentable in the USA under 35 USC §101, the new Russian law specifically includes patent protection for cell cultures and microorganisms. These may have been explicitly added because, under old Soviet regulations, patenting pharmaceuticals and medicines was considered antisocial. However, patent protection has not been extended to scientific theories and mathematical methods; methods of economic organization and management; conventional signs, schedules and rules; methods of mental health therapy; algorithms and computer programs; designs and schemes for the planning of installations, buildings and districts; decisions affecting only the external appearance of a product; topography of integral microcircuits; and varieties of flora and fauna (e.g. plant varieties and animal breeds), although the latter could be included if genetically engineered. Also not acceptable are inventions which are contrary to social interests or the principles of humanity and morality.

UTILITY MODELS. The Patent Law introduces a new concept, the utility model, as another form of IP, which is unknown in US legislation. While virtually anything can be patented as an invention, as long as it meets certain legal prerequisites, the subject matter of a utility model is narrowly circumscribed. Utility model certificates are provided only for technical devices, such as constructed objects, machines, instruments or industrial equipment. Processes, substances, cell cultures and microorganisms are explicitly excluded from being certified as utility models. The legal prerequisites for a utility model are less stringent than those for inventions: no level of inventiveness is required and, although novelty is required, the prior art standard is less strict than in the context of inventions. In addition, since there is no material examination, utility model certificates do not have any guarantee of validity and are issued under the full responsibility of the applicant with regard to such validity.

INDUSTRIAL DESIGNS. Like the US law, the Russian law defines industrial designs as an 'artistic solution of an article'. This expression is professionally used and understood in the fields of architecture and the artistic modelling of industrial designs. To 'solve an article' means to create an original, previously unknown model with a unique outward appearance, which allows the subsequent manufacturing of a series of articles in accordance with it. According to Russian law, a design enjoys legal protection if it is novel, original and industrially applicable. In contrast with patents, instead of being non-obvious, an industrial design must be 'original'. Although the definition of originality is rather vague, according to some commentators it lies in the creative nature of the aesthetic characteristics of the article, such as its unexpectedness or unusualness.

Patent application
Contrary to the US practice, Russia grants a patent to the first to file a patent application. The USA applies a first-to-invent standard, under which the first to file is not necessarily the party who is ultimately awarded a patent. The State Committee of the Russian Federation on Patents and Trademarks (*Rospatent*) is the sole office in Russia issuing patents. The Scientific Research Institute for State Patent Expertise (SRISPE) is designated to handle all applications and examine the patentability of the inventions on behalf of *Rospatent*.

The application for an invention must include: (i) a petition specifying the inventor; (ii) a description of the invention 'with fullness which is sufficient for effectuation'; (iii) the formula of the invention expressing its essence; (iv) sketches if necessary; (v) an abstract; and (vi) evidence that fees have been paid. Patent applications must be in Russian, although the supporting documents can be in another language at the time the application is submitted provided the translation is submitted within the following two months. As in the USA, the filing date is that on which the application is received at the SRISPE. It takes from six to 24 months from the date of the application to receive a definitive decision from SRISPE.

Patent examination

In contrast with the US law, the Russian law follows the practice of many European countries in dividing the patent examination into two steps: preliminary examination and substantive examination. Preliminary examination is aimed at ascertaining whether the documents comply with established formalities and whether the invention applied for is amongst the objects to which legal protection is granted. Substantive examination is an evaluation of the complete merits. Both of these examinations are appealable.

Term of patents

As in most industrial countries with first-to-file systems, a patent protection is granted for 20 years from the date the application is submitted to the SRISPE. The 20-year term is emerging as the international standard, adopted by various countries in Eastern Europe, Asia and Latin America. It is also the minimum patent term to which WTO members must adhere, according to Article 33 of the Uruguay Round's TRIPs Agreement. Thus, the Russian patent law is in this respect on a par with the world's standards.

Rights of inventors and owners

As in the USA, an inventor's authorship right (the right to be recognized as the inventor) is inalienable. However, other substantive rights flowing from the patent, as well as the right to obtain the patent itself, are transferable. Foreign investors should note, however, that in order to effect the assignment of patent rights under the Russian law, such an assignment should be handled by a licensed patent attorney and should be registered with SRISPE. Thus, it is always wise to check the records of SRISPE whenever one is dealing with a transfer of any patent rights (e.g. in cases of privatization, establishing joint ventures with Russian partners, takeovers, or other changes of patent holders). Otherwise, the patent may be useless. Foreign licensees should also note that Russian law goes further than US law in requiring that a license, even a non-exclusive one, must also be recorded. In cases of joint ventures with Russian partners, the foreign party should also understand that, unlike in US law, joint inventors may not preserve their rights by applying independently but must agree with each other with regard to their respective rights. Although each joint patent owner may individually use the invention, they may not assign or license it without the consent of all the owners. Therefore, it is advisable for foreign partners to reach an agreement regarding the future rights of each of the parties with respect to inventors' rights before establishing a joint venture or to make such an agreement a part of the corporate articles or by-laws. Furthermore, unlike in US law, the rights of employers and employees with regard to inventions are mainly prescribed by statute rather than by a labor contract. There is a presumption in Russian law that the employer owns the right to obtain the patent unless there is an agreement to the contrary. Thus, on the one hand, the inventions of Russian inventors employed by foreign companies or joint ventures belong to these companies. Analogously, as a rule, Russian inventors cannot legally assign their rights to inventions under development to foreign companies without the permission of their Russian employer. On the other hand, patents for the inventions of foreign inventors employed by joint ventures with Russian participation or by fully owned Russian companies will be recognized as belonging to the companies rather than to the individual inventors. Therefore, to avoid an abdication of their rights to inventions, foreign inventors should either work for Russian companies as independent contractors instead of being employed by them or explicitly stipulate their rights to any inventions in employment contracts.

Rights of foreigners
Foreign individuals or legal persons enjoy equal rights with Russian citizens according to international treaties of the Russian Federation or on the basis of reciprocity. If the rules of an international treaty contradict those of the Russian Patent Law, the former shall apply. However, despite the equal treatment of foreign and Russian inventors, commentators have observed that fees for a patent application in Russia are likely to be higher for foreign applicants as long as this practice is not prohibited by international treaty.

Thus, the new Patent Law of the Russian Federation provides adequate patent protection, comparable to international standards. Although the enforcement infrastructure is not yet sufficiently developed, the new Patent Law creates a promising legal framework that substantially increases the protection of the IPR of foreign investors in Russia.

PROTECTION OF RIGHTS ON PLANTS AND ANIMALS IN RUSSIA

The political changes in Russia have prompted legislative reforms in the area of protection of rights in plants and animals, which once was considered by the Soviets contrary to public order and morality. This section will focus on the Russian law dealing with the protection of plant varieties and animal breeds.

Protection under the Patent Law

The Patent Law of the Russian Federation, as mentioned above, explicitly excludes plant varieties and animal breeds from patent protection, even if the qualifications of novelty, inventiveness and industrial applicability are met. However, such an exclusion is not a unique feature of the Russian law. This exclusion provision is uniform with section 53(b) of the European Patent Convention of 1973 (EPC) which declares that European patents shall not be granted with respect to plant or animal varieties or essentially biological processes for the production of plants or animals. Commentators usually give three reasons why plant and animal varieties are excluded from patent protection:

1. It is virtually impossible to give an accurate reproducible description of a plant or animal variety.
2. Most plant and animal varieties bred by conventional techniques (controlled pollination and selection for plant breeding; and selective breeding and cross-breeding, either through natural mating or artificial insemination, for animal breeding) lack inventiveness.
3. Plants, plant varieties, animals and animal varieties do not need to be under patent protection *per se* because they are usually the subject matter of protection under the various national plant variety acts, such as the Law of Russian Federation on Selection Achievements, which was adopted on 6 August 1993.

Although the Patent Law of the Russian Federation excludes patent protection for plant varieties and animal breeds, it does provide protection for microbiological processes and their products (such as strains of microorganisms, cell cultures of plants and animals and the application of previously known structures, means, substances and strains for a new purpose) and, arguably, plant and animal varieties produced by genetic engineering techniques (including recombinant DNA techniques).

Protection under the Law of the Russian Federation on Selection Achievements

The Law of the Russian Federation on Selection Achievements offers legal protection for plant varieties and animal breeds, which are protected by a special patent on the selection achievement certifying the exclusive right of the patentee to the use of such selection achievement. This law is modeled on the 1991 version of the International Convention for the Protection of New Varieties of Plants (UPOV), which was signed by the member countries in 1961 and revised in 1972, 1978 and 1991. In contrast with this Convention, which is aimed to protect plant varieties, the Russian Law on Selection Achievements addresses the legal protection of both plant varieties and animal breeds.

Plant varieties
The Russian Law on Selection Achievements defines a variety as a plant grouping that, irrespective of whether it is eligible for protection, is defined by features characterizing a given genotype or combination of genotypes, and is distinguished from any other plant grouping of the same botanical taxon by one or more features. This definition is very similar to that of the 1991 UPOV Convention. A variety may be one or several plants or parts of plants if such parts can be used for the reproduction of entire plants of the variety. The law further explicitly designates such protected categories of a variety as a clone, line, population and hybrid of the first generation. Nevertheless, the law apparently considers that new types of varieties such as multilines or multiclones do not come within this definition. It is not clear why the Russian legislators chose to adopt this narrow definitive approach, which was revoked by the 1978 Revision of the UPOV Convention as over-restrictive. Furthermore, in compliance with the 1991 UPOV Convention and international obligations of the Russian Federation, the Russian State Commission adopted a list of botanical and zoological genera and species for which protection should be available.

Animal breeds
Under the Russian Law on Selection Achievements, a breed is defined as an animal grouping that, irrespective of whether it is eligible for protection, has genetically conditioned biological and morphological particularities and features some of which are typical for such groupings, and distinguish it from other animal groupings. A breed can be male or female, with specimen or breeding material. The law distinguishes such protected categories of breeds as type and cross lines. The law defines 'breeding material' as a breeding animal, its gametes or zygotes (embryos). Furthermore, the law differentiates two types of animals: breeding animals and commodity animals. 'Breeding animals' are defined as those destined for the reproduction of a breed, while 'commodity animals' are those used for purposes other than the reproduction of a breed.

The criteria for protecting a selection achievement
For selection achievement (successfully accomplished selection) to be capable of protection, they must be novel, distinct, uniform and stable. A variety or breed is considered 'novel', if on the date of filing of the application for a patent, the seed or the breeding material has not been sold or otherwise transferred to others, by or with the consent of the breeder or his legal successor, for purposes of exploitation of the selection achievement at the following times:

- in the territory of the Russian Federation, earlier than one year before the filing date;
- in the territory of another state, earlier than four years;
- in the case of vines, ornamental trees, fruit trees and forest trees, earlier than six years before the said date.

In contrast with patent law, the meaning of 'novelty' in this act is commercial novelty rather then technical novelty, which refers to something that is innovative and is not known from prior art. A selection achievement lacks novelty if it was commercialized by the breeder prior to the filing date.

The selection achievement applied for must be clearly 'distinguishable' (by one or more important morphological, physiological or other characteristics) from any other selection achievement, whose existence is a matter of common knowledge at the moment the application is filed. A selection achievement may be considered commonly known when it is found in official catalogs or information funds, or when it has been described in a precise manner in any of the publications. The filing of an application for a patent on a selection achievement or for the permission to use it also makes the selection achievement commonly known from the date of the filing of the application if the patent was actually issued or the selection achievement was allowed to be used. The meaning of 'distinctness' in this act is 'technical novelty', which is based on the concept of common knowledge analogous to the prior-art concept in patent law. Thus, the criteria of 'novelty' and 'distinctness' as used in the act together have in a sense the same meaning as the criterion of 'novelty' used in patent law.

The plants of a given variety and the animals of a given breed must be sufficiently 'uniform' in their features, taking into account the individual deviations that may take place in connection with the particularities of the propagation. The requirement of uniformity, however, limits the amount of divergence allowed and so makes ineligible for protection some of the cross-pollinated plant species that commonly exhibit diversity in character within a variety.

Finally, for a selection achievement to be capable of protection, it must be 'stable'. A selection achievement is considered stable if its basic features remain unchanged after repeated propagation or, in the case of a particular cycle of propagation, at the end of each such cycle. Although stability poses no problem with asexually propagated plant varieties, it can pose a great problem with sexually reproduced varieties, because cross-pollination can lead to a shift in type with loss of the important characteristics described in the application.

Scope of protection
The selection achievement for which a patent is granted by the State Commission is registered in the State Register of Protected Selection

Achievements. The patent is valid for 30 years from the registration date or, for varieties of vines, ornamental trees, fruit trees and forest trees, including their rootstocks, for 35 years. The scope of the legal protection granted by the patent for a selection achievement is determined by the totality of the essential features fixed in the description of the selection achievement.

A patent for a selection achievement gives the patentee the exclusive right to the use of the selection achievement. If any one wants to produce or reproduce the protected selection achievement, condition it for subsequent propagation, offer it for sale, sell it, or market it in other ways, export or import it, or store it for the aforementioned purposes, they must obtain from the patentee a license to do so. The right of patentee is also extended to vegetable material produced from seed, or to commodity animals produced from breeding animals that are commercialized without the authorization of the patent owner. The monopoly of the patentee over the protected selection achievement, however, does not extend to acts performed for personal, non-commercial or experimental purposes, or for the purpose of creating other varieties and breeds. Furthermore, the exclusive right of the patentee does not extend to the following acts performed with the protected selection achievement: (i) the use of vegetable material, obtained in an enterprise, for two years for propagating purposes on the facilities of this enterprise; (ii) the reproduction of commodity animals to be used in a given enterprise; and (iii) any acts involving seeds, vegetable material, breeding material and commodity animals that have been commercialized or agreed to be commercialized by the patentee.

License agreement

The patentee can transfer the right to use the selection achievement to another person by entering into an exclusive or non-exclusive license agreement. When the patentee grants an exclusive right to use the selection achievement, the licensee, within the limits provided by the agreement, takes all the profit and is solely entitled to the selection achievement, even to the exclusion of the patentee. By contrast, when a license is non-exclusive, the patentee may grant similar rights to others or may use the selection achievement itself. An exclusive license is effective only after registration with the State Commission. The licensee may not transfer the license or grant sublicenses to a third party if this was not provided in the license agreement. In the case of an infringement of the rights of the patentee, the licensee may file a lawsuit on its own behalf as though the action was initiated by the patentee.

The patentee may also grant an open, non-exclusive license by publishing in the official bulletin of the State Commission a statement that any person has the right to use the selection achievement upon the payment of the specified compensation if the patentee is notified of such use. To encourage patentees to grant such open licenses, the law has halved the fee required to be paid by patentees to maintain the patent's validity. In certain circumstances, such as when a patentee refuses another person the right to produce or raise seeds or breeding material and there are no legitimate reasons to prevent the patentee from granting this person the right, the State Commission may issue a compulsory license. Upon the issuance of a compulsory license, which cannot be longer than four years, the State Commission determines how much the holder of the compulsory license must pay to the patentee. The decision of the State Commission on granting the compulsory license, however, may be appealed against in court.

Procedure for the grant of patents

An application for a patent on a selection achievement has to be filed with the State Commission of the Russian Federation for Examining and Protecting Selection Achievements by the breeder or its legal successor. If the selection achievement was grown, created or revealed by somebody in the course of their duty to their employer and during the course of employment, the right to file an application for the granting of a patent belongs to the employer, unless otherwise provided in an agreement between the breeder and the employer. The application may be filed through an intermediary who is entitled on the basis of a power of attorney to conduct the affairs in relation to the receipt of the patent, which is particularly important for out-of-state companies. When the applicant is the employer, it must corroborate the existence of an agreement with the author of the selection achievement. All documents and the application should be submitted in Russian. The priority of the selection achievement is determined by the date of receipt of the application by the State Commission.

The application must contain a request for the granting of a patent on the selection achievement, a completed questionnaire on the selection achievement, a document confirming the payment of the required fees, and a name for the selection achievement. The name of the selection achievement must enable the selection achievement to be identified, and must be brief and distinguishable from the names of existing selection achievements of the same or of closely related botanical or zoological species. It may not consist solely of figures, nor mislead with regard to the particularities, origin and meaning of the selection achievement, or with regard to the identity of the breeder, and must not contradict the principles of humanity and morality.

The preliminary examination of the application takes one month, during the course of which the priority date is determined and the necessary documents are examined to check that they conform with requirements. If everything is in order, the applicant is informed that the application has been accepted and details of the application are published in an official bulletin. In the six months following this publication, the State Commission examines the novelty, distinctness, uniformity and stability of the selection achievement. The applicant must submit the necessary quantity of seeds and breeding material for examination by the State Commission. If the selection achievement conforms to the criteria for protecting a selection achievement, and its name conforms to the conditions discussed above, the State Commission decides to grant a patent and draws up a description of the selection achievement.

The rights of the author of a selection achievement
When the patentee is not the 'author' of the selection achievement, i.e. the person by whose creative labor the selection was grown, created or revealed, it must pay the author for the use of the selection achievement. The author of a selection achievement has the right to receive, from the patentee, remuneration for the use of the selection achievement during the term of the patent. The size and conditions of payment are determined in an agreement between the patentee and the author, but the remuneration must not be less than 2% of the annual income received by the patentee from the use of the selection achievements, including the income from the sale of licenses.

An author of a selection achievement is entitled to an author's certificate, which verifies the authorship, as well as the right of the author to receive remuneration from the patentee for the use of the selection achievement. An author's certificate is issued by the State Commission to every author who is not the patentee. An entity cannot be an author of a selection achievement. All disputes about authorship can be settled in court.

PRACTICAL RECOMMENDATIONS

With the collapse of the Soviet Union, Russian lawmakers have managed to remedy most of the problems of protecting IPR, and the modern Russian IP laws now provide adequate protection of the interests and rights of foreign investors doing business in Russia.

Following the dissolution of the Soviet Union, the Russian Federation as a successor state assumed the treaty obligations of the USSR, including those under the Agreement on Trade Relations Between the United States and the Union of Soviet Socialist Republics, which was signed by the countries on 16 June 1992. This bilateral agreement, which was negotiated and signed by Mikhail Gorbachev just before the disintegration of the Soviet Union, provides Russia with 'the most favored nation status' in exchange for commitments on IP protection. The agreement commits Russia to adhere to the Paris

Convention, the Universal Copyright Convention and the Berne Convention. Russia has complied with these and most other major provisions of the USA–Russian bilateral agreement by passing the domestic legislation on IP protection described above.

Nevertheless, residual problems arising from previous legislation and undeveloped enforcement mechanisms weaken the practical protection and therefore create various obstacles to foreign participation in joint ventures with Russian partners. Moreover, in the light of frequently changing legislation and flourishing corruption among Russian officials, it is irrational to rely exclusively on the letter of the law in protecting one's rights. Instead, self-help measures might be more practical. Some general recommended self-help measures which can enhance the protection of IPR in Russia are discussed below.

First, the emphasis in protecting one's IPR should shift from the legislative level to the individual level with strong contractual provisions. It is much easier to resolve all ambiguities between parties at the negotiation stage than to rely on the default rules for resolving disputes which are provided by legislation. Contractual provisions should not only clarify the parties' mutual rights and obligations, but also provide guarantees for enforcement of such rights and remedies in the case of a breach of contract. For instance, providing for liquidated damages in the contract is a more feasible and practical alternative to legislative sanctions. To avoid uncertainty, the parties should specify the choice of law and forum, to protect against unfavorable contract interpretation. An arbitration clause, which is recommended to be included in any contract, should be very specific. The parties should stipulate the location and language of the arbitration, the number of arbitrators and the means of their appointment, the choice of law governing the substance of the disputes, and the rules that will govern the arbitration.

Second, as some commentators have observed, until effective enforcement mechanisms become a reality in Russia, a strong presence in the Russian market is often the critical factor. A 'presence' can be established through franchise or exclusive dealership distribution systems. Although a foreign company may be able to protect its rights abroad, it is much easier for franchisees and dealers located inside Russia to identify and thwart infringement of the company's IPR in the country, as well as to represent the company in Russian judicial and administrative infringement proceedings.

Third, to limit potential infringement of IPR, foreign business people should avoid or minimize distribution of especially sensitive products. To minimize the circle of individuals having access to sensitive information, one should segregate license provisions (containing a substantial amount of technical information) from primary joint ventures or technology transfer contracts, as well as from any other contracts between the parties. The parties should draft reasonably worded confidentiality provisions in terms of duration and other factors to enhance their enforceability. Due to the corruption of the state, foreign businesspersons should disclose the minimum information necessary to government patent agencies and arbitration courts. Although registration is recommended, in certain situations it may be better not to register an intellectual product, such as a computer program, rather than risk subjecting the product to potential abuses by government officials.

Fourth, a licensing arrangement is generally preferable to a joint venture agreement. Licensing offers flexibility in technology choice and a better opportunity to negotiate. Technology license agreements enable a licensor to take advantage of the Russian market without risking capital in an unpredictable foreign climate. However, many Russian manufacturing sectors at present lack sophisticated industrial technology and know-how, and merely licensing technology may not be sufficient for the Russians to adopt such foreign technology to the domestic production process successfully. Thus, even simple license arrangements with the Russians often contain comprehensive provisions on technical assistance and training, and turnkey ventures and joint ventures have become popular. One strong incentive for a Russian party to enter into a joint venture is that it need not pay in cash for the transferred technology but may instead contribute various in-kind capital, such as raw materials, existing industrial facilities and structures, existing patents, licenses and permits, trademarks or other IP, and participate in the export sales of goods produced with such technology, collecting much-needed convertible currency.

Finally, several practical suggestions are given regarding how a foreign investor should behave while doing business in Russia.

1. For cultural reasons, one should treat a Russian partner equally. Indeed, very often a concern about equal distribution of profits is much more an issue for a Russian negotiator (who is afraid he will not be treated fairly) than anything else.
2. It is always wise to establish good personal relations with the key persons on the Russian side of the partnership. If possible, involve Russians on your side when negotiating with other Russians.
3. Try to put everything in writing.
4. The support of local municipalities is always helpful and should never be ignored. One should establish amicable relations with the local authorities, in order to avoid unnecessary obstacles in the future.

Finally, foreign companies should supplement these basic precautions with professional legal advice. Despite the general legislative advances detailed in this chapter, the Russian market is still undergoing rapid change. This market still operates according to unwritten principles that frequently contradict those recently legislated. Competent legal counsel can provide timely insight and ensure a higher probability of success for foreign companies doing business in Russia.

Andean Pact Countries of Latin America

12

Walter R. Jaffe and Elinor Arteaga-Marcano

CONICIT, Edf. Malploca, Av. Principal de los Cortijos, Caracas, Venezuela

INTRODUCTION

The Andean Pact, or Pact of Cartagena (*Acuerdo de Cartagena*), named after the place were the pact was signed, was created in 1969 and initially comprised Bolivia, Colombia, Chile, Ecuador and Peru. Venezuela entered in 1973 and Chile left in 1976 (IFEDEC, 1987). Its stated objective is the balanced and harmonious development of its member countries through economic integration. Towards this end, its most important instruments are the harmonization of economic and social policies, common industrial policies, liberalization of trade within the group, a common external tariff and communication infrastructure development.

The maximum political authority of the Andean Pact is the Commission, formed by national representatives. The commission normally meets three times a year. Its decisions are numbered sequentially and have the force of law in each member country. The technical and executive body of the Pact is the *Junta del Acuerdo de Cartagena* (JUNAC), which is supported by an administrative and technical office located in Lima, Peru. The Junta has three members named by the Commission for a three-year term. Additional important structures within the Andean Pact framework are the Andean Parliament and Court and the Andean Finance Corporation (*Corporacion Andina de Fomento*), among others.

The Andean Pact was launched as a political and economic development mechanism. Each country sought to develop a local industry using tariff and other barriers to protect it in its initial 'juvenile' phase.

Seeking to develop an industrial base, governments intervened heavily in the economy through planning and direct investment. This basic policy was contradictory to the objectives of the economic integration efforts leading to a progressive slowing down and paralysis of the initially dynamic Andean Pact.

In its first 20 years of existence the Pact was not able to implement any of its ambitious integration programs on a significant scale. The most difficult proposition was the establishment of common industrial policies in key sectors, like automobiles, which sought to establish the development of coordinated and complementary industrial capabilities. So the manufacture of specific types of cars and components were assigned to different countries. After years of acrimonious negotiations, these attempts were abandoned in the 1980s.

Intraregional trade grew modestly in the 1970s, increasing from 2.1% of the total foreign trade of the Andean countries in 1970 to 4.6% in 1979. The economic crisis of the 1980s affected each member country's integration efforts. The initial response to the crisis produced a first generation of structural adjustment programs which had an important impact on trade, and in particular, on intraregional trade, which declined by about 40% and then stagnated (IFEDEC, 1987).

Later, the new paradigm of open economies and free trade took hold in all Andean Pact countries, producing a reassessment and reorientation of the regional economic integration efforts. The key event was the bilateral decision of Colombia and Venezuela, taken in 1989, to completely liberalize their trade within a year. This produced an important rebirth of the Andean Pact; this time it basically centered on intraregional free trade and establishment of a common external tariff.

Trade within the member countries has grown exponentially and has reached about US$3 billion per year in 1994. Its annual growth in the period 1990–1994 was on average 31% (authors' calculation based on information given in the JUNAC WWW site at http://ekeko.rcp.net.pe/junac/UIO06.htm). Trade is now essentially free between Colombia, Venezuela, Ecuador and Bolivia: Peru still having doubts about joining the chosen path. Cross-investment, principally between Venezuela and Colombia, has increased hugely. Many local companies have set up subsidiaries in neighboring countries. This success has revitalized other areas within the Andean Pact, one being the enactment of trade-related regional legislation.

DEVELOPMENT OF IPR PROTECTION IN THE ANDEAN PACT COUNTRIES

The most representative policy of the Andean Pact in its first period of existence was the famous Decision 24 of December 1970 which established a common regime for the treatment of foreign capital and brands, patents, licenses and royalties (JUNAC, 1982). It heavily regulated foreign investment in the member countries, providing strong support to national capital. Whole economic sectors like financial services and banking, for example, were reserved for national capital. Foreign investment as well as capital export had to be authorized by government. The progressive conversion of foreign companies into national ones was foreseen.

The widely held perception of abuses in the licensing of foreign technology to local companies also led to standards designed to strengthen the bargaining position of national companies, vis-à-vis foreign providers of technology. Licenses had to be approved by government and had to conform to certain standards. Restrictions for the licensee on use of the technology and of the products manufactured with it were forbidden. Member countries were instructed to set up specialized intellectual property (IP) offices.

In 1974 standards for granting and managing patents and brands were established through Decision 85 (JUNAC, 1982). This decision developed the basic tenets of Decision 24 in the area of IP. Patentability requirements and administrative procedures to grant them were defined. The protection granted was generally weak, in response to the idea that this system benefited basically foreign companies. The period of patent protection was, for example, ten years and strong compulsory licensing possibilities were established.

In the current open-economies phase of the Andean Pact Decision 24, its related standards were abandoned and substituted by new regulations allowing foreign investments and international movement of capital. In the area of IP, stronger protection of patents and brands was provided, and new areas, such as industrial secrets and denominations of origin, were opened. This new trend started with Decision 311 in 1991 and Decision 313 in 1992. As a result, the IP regime of the Andean Pact countries conforms closely to standard international practice (SELA, 1994).

The current intellectual property rights (IPR) policy of the Andean Pact countries includes a common patent and plant breeders' rights (PBR) system, expressed in Decisions 344 and 345 and published in the *Andean Gazette* no. 142 of 29 October 1993. More recently, a common policy regulating access to genetic resources was agreed upon, which has some IPR implications. It is contained in Decision 391 published in *Gazette* no. 213 in July 1996.

The Andean patent regime, defined by Decision 344, follows closely the World Intellectual Property Organization (WIPO) directives. It permits the patenting of all products or processes in all fields of technology if they are new, inventive and can be applied industrially (JUNAC, 1994a). Animal species and races and the biological procedures needed to obtain them are explicitly excluded, together with certain other categories. Microorganisms and other living beings (e.g. plant species) could be patented. Article 13 specifies that an invention related to living beings and materials must be deposited in an institution authorized by the national authorities, and include materials which will be part of the description of the patent document.

Decision 345 establishes a common PBR system based on the UPOV model. The potential contradiction between two different forms of protection of plant species or varieties has yet to be resolved.

The supranational legislation is applied in each country. In the case of Decision 345, national by-laws are required. The status of the actions taken by the member countries in this regard to December 1996 are presented in Table 12.1. The validity of the supranational legislation in the member countries has been challenged in the courts, which have until now upheld it. For example, several rulings by the Venezuelan Supreme Court have tacitly recognized the validity of Decision 344.

DESCRIPTION OF IPR RELATED TO AGRICULTURE IN THE ANDEAN PACT COUNTRIES

Plant breeders' rights

The IPR legislation most directly related to agriculture is the PBR system. In the case of the Andean group, its principal objective is to recognize and guarantee the protection of the PBR over new plant varieties through the issuing of a certificate valid in all Andean Pact countries.

Decision 344 considers that a breeder has created a new variety when he or she has applied scientific knowledge to obtain a homogeneous, distinct and stable variety and when he has given it a generic denomination (JUNAC, 1994b). Only varieties created that satisfy these criteria can be protected. That is, discoveries are excluded. This gives

Table 12.1. National by-laws of Decision 345 of the Andean Pact (status as at December 1996).

Country	Official act and date
Bolivia	By-law in project stage
Colombia	Decree no. 533, March 1994
Ecuador	Executive decree 3708 published in *Official Register* no. 925, April 1996
Peru	Supreme Decree 008–96-ITINCI published in *Gazette El Peruano*, May 1996
Venezuela	By-law in project stage

exclusive rights to the creators for a period of 15–25 years depending on the species. The breeder has to deposit a living sample of the variety.

Decision 345 makes it possible to extend the rights to essentially derived varieties. A variety is considered to be essentially derived when a variety, although distinguishable from other varieties, is predominantly derived from another variety and also retains essential characteristics which result from the genotype of that other variety. Therefore, the breeder is not harmed by whoever reserves and plants, for his/her own use or sells as prime material or food, the product obtained from the cultivation. Exceptions include the commercial use of multiplication, reproduction and propagation materials, including whole plants and their parts, of fruit, ornamental and forestry species.

The breeder will have provisional protection dating from the presentation of the claim to the granting of the certificate. The Decision also gives priority rights over any other request, during a 12-month period, seeking the protection of the same variety in any other Andean Pact country. The certificate holder grants licenses for the protected variety.

Decision 345 permits the use of protected varieties for noncommercial purposes, for experimental uses and for the obtention of new varieties. Its implementation in each country requires a national by-law. Colombia, Peru and Ecuador have implemented regulations while Bolivia and Venezuela have only proposed regulations.

As established in the Decision, each member country has to create a national registry of protected plant varieties. As of December 1996, only Venezuela has not followed this obligation.

Additionally, JUNAC maintains a regional registry of protected varieties. Created by Decision 345, this registry is governed by a Regional Committee for the Protection of Plant Varieties. The Decision establishes directives that harmonize the procedures, laboratory tests, deposits and cultivation of samples necessary for the registration of a variety, define technical criteria for distinguishing between varieties, so as to determine the minimum number of characters which have to change to be able to determine if a variety is different from another and analyze matters related to the protection of essentially derived varieties and propose norms for its Andean group. Up to December 1996, this Committee has met four times (JUNAC, 1996a).

Colombia is the only country of the Andean Group which is a UPOV member. In September 1996, she became party to the UPOV Convention and to the 1978 Act (see WWW page at http:wipo.org.ratific/t-upov.htm).

Access to genetic resources

The Andean Group countries, led by Colombia, have significantly advanced the operation and implementation of the Biodiversity Treaty.

Most countries in the world signed this treaty at the Earth Summit of Rio in 1992. Decision 345 declared the intention to regulate access to genetic resources. This led to the approval in July 1996 of Decision 391, which regulates the access and use of genetic resources native to the member countries (JUNAC, 1996b). In this way, a link between PBR and access to genetic resources was established. The link was created by Decision 391 (JUNAC, 1996a) and reinforced by the intended merger of the Regional Committee for the Protection of Plant Varieties and the Andean Committee of Genetic Resources. This last decision could therefore have important practical IPR implications.

Decision 391 essentially establishes a system for authorizing the access and use of native genetic resources. National authorities will contract with interested parties regulating the conditions of access and of use of genetic resources, including the sharing of benefits (JUNAC, 1996c). An Andean Committee of Genetic Resources will serve as an information clearing house, technical advisory and harmonizing body for the authorization process.

The definition of genetic resources adopted in this decision ('all biological material which contains genetic information of real or potential value') includes plant varieties. Therefore, the potential for overlap and conflict exists in PBR.

IMPACT OF IPR ON AGRICULTURE IN THE ANDEAN PACT COUNTRIES

The rapid privatization of biological technologies, traditionally mostly in the public domain, brought about by the development of biotechnology, has increased the importance of IPR in many sectors, particularly in agriculture. Special IPR systems for agriculture, i.e. PBR, are increasingly complemented by the application and use of more traditional ones, such as patents. But many political and practical issues remain to be solved for an extensive and effective protection of IPR in this sector.

Theoretically, IPR could impact agriculture in several ways (Jaffe and van Wijk, 1995). The first and most direct would be the acceleration of the rate of technological innovation in agriculture, which is, after all, the economical justification of an IPR system. This could occur both in agricultural practices directly, or indirectly in industries which produce inputs for or process prime materials from agriculture. The basic condition is that technologies could be protected effectively. Changes in R&D investments and distributions are indirect indicators of trends in this respect.

The structure of agricultural production could also be affected by IPR, if they determine the access to and use of technologies. Differences in the prices and costs of protected or unprotected technologies, which have an effect on the competitiveness of different production systems or industries depending on the scale of production, could favor certain systems or industries over others. The most common worry here is the perceived negative effects of IPR on traditional, small-scale production systems.

Finally, IPR in agriculture could have an impact on the relationship of this sector with other ones. One possibility would be a strengthening of agriculture against consumers for example, that is an increase in the cost of agricultural products for consumers.

In general, Andean group countries have weak local or national innovation systems. Particularly in agricultural industries where technologies can be protected, there are only very limited local capabilities for generating technologies. This means that IPR in these countries will be more important for the international access to needed technologies and products than for the stimulation of local technological innovation (van Wijk and Jaffe, 1996). A good example of this is the case of the flower industry in Colombia. This industry, which in the last 20 years has grown into one of the most important worldwide, has an international reputation of not respecting IPR of flower varieties. Threats of punitive restrictions to access markets, principally the crucial American one, was the most important reasons for the relatively rapid enactment of Decision 345, a legislative track which was perceived as easier than a national law. This generated a conflict between the defenders of PBR in Colombia, the flower and seed industry, the Trade Ministry, Congress, non-governmental organizations and some political parties, which were in favor of a national, much more restrictive, law.

Plant breeding has been mainly a public sector activity in the Andean group countries and is concentrated in the National Agricultural Research Institutes (NARIs) of each country, as well as in the two international centers located in Colombia and Peru. The most important programs have been in maize (corn), rice, sorghum, potatoes, sugarcane and beans. Some breeding has also been done by universities. In general, breeding consists of adapting foreign materials, commonly from the international centers in the case of food crops, to local conditions through back-crossing. The only significant private breeding activity is that carried out by coffee growers in Colombia. Some local industries have had limited experience in adaptive breeding of specific crops, like the example of sorghum in Venezuela and sugarcane in Colombia. It is to be expected that the existence of PBR, coupled with effective enforcement, should stimulate the investment of private companies in plant breeding.

The seed industry in the Andean group countries generally relies on foundation seed produced by the NARIs to reproduce and commercialize. The public sector varieties, until the recent approval of the PBR system, were inadequately protected so that the NARIs could not profit from them. The enforcement of PBR could result in an important source of revenues for these institutes, as the experience of Argentina shows (Jaffe and van Wijk, 1995). In the present context of strong budgetary restrictions which all these institutions face, this could be an important element in their future prospects and roles.

Multinational seed companies have had a strong, and in many cases dominant, presence in the Andean group countries for many years. In particular, the larger markets of Colombia and Venezuela have been attractive for hybrid maize, sorghum, cotton and vegetable seed companies. They generally do only adaptive breeding, if any, in the host countries. The lack, until recently, of adequate IPR protection in the Andean group countries has not been a factor in the decision to locate seed production facilities in these countries. Market size and general business climate are generally the determining reasons for these decisions. On the other hand, these companies generally deal with hybrids, which mostly do not need legal IPR protection.

IPR related to agriculture, and specifically PBR, have not been in place for long enough to have had any real impact on agriculture in the Andean group countries. No effective enforcement structures existed until December 1996, given that the most some of the countries have advanced is in the enactment of legislative measures and the setting up of public-sector organizations to manage and enforce them. But, as the experience of Argentina clearly shows, effective enforcement of PBR requires that breeders organize themselves to this end, and this has not yet happened in any of the Andean group countries.

REFERENCES

IFEDEC (1987) *La Decision, Aportes para la Integracion Latinoamericana.* Coleccion Seminarios, Caracas.

Jaffe, W.R. and van Wijk, J. (1995) The impact of plant breeders rights in developing countries. In: *Special Programme Biotechnology and Development Cooperation.* Directorate General International Cooperation, Ministry of Foreign Affairs, The Netherlands.

JUNAC (1982) *Ordenamiento Juridica del Acuerdo de Cartagena*, Tomo I, Decisiones 1–90. Lima, Peru.

JUNAC (1994a) *Decision 344. Regimen Comun sobre Propiedad Industrial.* Gaceta Oficial de la Republica de Venezuela no. 4676, Extraordinario. Caracas, Venezuela.

JUNAC (1994b) *Decision 345. Regimen Comun de Proteccion a los Derechos de los Obtentores de Variedades Vegetales.* Gaceta Oficial de la Republica de Venezuela no. 4676, Extraordinario. Caracas, Venezuela.

JUNAC (1996a) *Informe de la Cuarta Reunion del Comite Subregional para la Proteccion de Variedades Vegetales.* Quito, Ecuador.

JUNAC (1996b) *Decision 391. Regimen Comun sobre Acceso a los Recursos Geneticos.* Gaceta Oficial del Acuerdo de Cartagena no. 213. Lima, Peru.

JUNAC (1996c) *Resoluciones 414 y 415. Adopcion de Modelos Referenciales de Solicitud de Acceso y Contrato de Acceso a los Recursos Geneticos.* Gaceta Oficial del Acuerdo de Cartagena no. 217. Lima, Peru.

SELA (1994) *Cambios y Tendencias Recientes en la Legislacion de Propiedad Industrial a Nivel Regional e Internacional.* Documento preparado por la Oficina Internacional de la OMPI. SP/IV/Foro/PI/DI no. 1, Caracas.

Van Wijk, J. and Jaffe, W.R. (1996) Plant breeders' rights in Latin America. The effect of transfer of foreign plant varieties. *Science Communication* 17(3), 338–356.

13

Costa Rica

Silvia Salazar

PO Box 8-5750-1000, San José, Costa Rica

CURRENT STATUS AND CHANGES OVER THE LAST DECADE

In Costa Rica, intellectual property rights (IPR) are protected in the Constitution. Article 47 of the Constitution establishes that, according to law, every author, inventor or producer will be granted a temporary, exclusive right in his or her creation, invention, trademark and commercial name. Based on that promulgation, Costa Rica has implemented a series of laws and subscribed to many international conventions related to IPR protection. The rights that are protected include patents, utility models, industrial models and designs, trademarks, commercial names, origin denominations and copyrights. IPR laws in Costa Rica are very strong and are on a par with human rights. In the field of industrial property, Costa Rican IPR laws regulate patents, models and industrial designs, utility models, trademarks, commercial names and advertizing. The Patent Law dates from 1983. Other important laws related to IPR include the Seeds Law and the Wildlife Law (Palacios and Salazar, 1995).

Costa Rica is a member of the World Intellectual Property Organization (WIPO) and continues to receive its technical assistance and training. The country is also a member of the 1916 Buenos Aires Convention and, very recently, the Paris Convention.

As of 1970, trademarks are protected in Costa Rica by the Central American Convention. Consequently, trademark laws in member countries, Guatemala, El Salvador, Nicaragua and Costa Rica, are the same. While trade secrets are not specifically mentioned in Costa Rican laws, some references can be found in both the Criminal and Labor Codes.

The 1982 copyright law has also undergone many changes and reforms. Most of these changes implemented the Berne Convention, Geneva Convention, Rome Convention and the Phonogram Convention. Costa Rican intellectual property (IP) laws are summarized in Table 13.1.

Piracy is not, and never has been, a great problem in Costa Rica (R. Sherwood, personal communication, 1996) even though the Patent Law and the Copyright Law were promulgated only 15 years ago. After Costa Rica signed the General Agreement on Tariffs and Trade (GATT; now the World Trade Organization (WTO)), the country recognized that substantial changes were needed in the existing IPR laws.

As in most developing countries, IPR protection has never been well known or studied much in Costa Rica. Until the 1980s, IPR laws dated from the last century; for example, the Nicaraguan Patent Law dated from 1899. The laws were there but no one followed them. However, inspired by a critical movement in the 1970s, the situation changed in the 1980s. Big changes were also made in most Latin American countries' IPR systems – especially patents. Studies and papers began to criticize the distortions and problems associated with patents produced in Latin American economies (SELA, 1988). This situation continued in a majority of Latin America before and during the GATT negotiations. Some of the characteristics of these systems

Table 13.1. Costa Rican intellectual property system.

Category	Components
General	Constitution
International conventions	Paris Convention
	Berne Convention
	Buenos Aires Convention (1910)
	Washington Convention (1946)
	Geneva Convention (1952)
	Rome Convention (1961)
	Geneva Convention (1971)
International agreements	Uruguay Round, GATT (WTO)
Free trade agreements	Costa Rica–Mexico (1994)
Industrial property	Patent Law
	Regulations to the Patent Law
	Seeds Law
	Wildlife Conservation Law
	Criminal Code
	Labor Code
	Central American Convention on Industrial Property
Copyright	Copyright Law
	Regulations to the Copyright Law

included: (i) weak protection in some fields and a lack of protection in others, with the latter especially found in pharmaceutical patents and agrochemical areas; (ii) short patent terms; and (iii) lack of IPR enforcement.

The situation changed greatly after the announcement of the Agreement on Trade-Related Aspects of Intellectual Property Rights, including Trade in Counterfeit Goods (TRIPs). Many Latin American countries, including Costa Rica, soon became members of the WTO and had to modify their existing IPR laws. Even today, there is a debate as to the real reasons why these laws were changed.

There has not been a real and detailed debate on IPR protection in Costa Rica. Most of the changes in the laws that have occurred, especially the Copyright Law, are not the products of studies or debates, but, instead, are in response to specific demands from developed countries like the USA. In a global economy with open markets and free trade agreements, the IPR rules will also have to be changed to remain competitive (Sherwood, 1990). Unfortunately, in Costa Rica these changes have been made without considering the impact, positive or negative, of those changes on the country's socioeconomic development. With clear rules now articulated under TRIPs and with no possibility of reverting back to the old laws, Costa Rica should realize that IPR systems must be conceived in accordance with a chosen economic development model. Unfortunately, Costa Rica is making only patchwork changes which reflect neither a policy nor a strategy for the social and economic development of the country.

The Costa Rican patent system is very different from the American patent system. Unlike the USA's right to exclude others from using the patented invention system, the Costa Rican patent owner is granted the right to exclusively exploit the patent and give licenses to third parties. The invention can be a product, a machine, a tool or a process, and any improvements can also be patented. It is important to mention that Costa Rican law excludes some inventions or fields from patentable subject matter. These exclusions are: discoveries, scientific theories, mathematical methods and software. Also excluded are plant varieties, animals and the biological processes used to obtain them, microbiological processes and their products, plans and principles, economic or business methods, original methods, intellectual activities, game rules, therapeutic and surgical methods, methods of diagnosis applicable to human beings and animals, and inventions contrary to the public health, security, public order and morals.

In accordance with usual patenting principles, the three requisites for patentability are novelty, non-obviousness and utility. Although most Costa Rican patents are granted for 12 years from the date of issue, some patents are granted only for one year. These include pharmaceuticals, therapeutic articles and substances, beverages, food, fertilizers and agrochemicals. Practically speaking, these areas are not generally patented due to the short patent duration. Patents from abroad are also granted protection but cannot exceed 12 years. Other important characteristics of the Costa Rican patent system are:

- the obligation to exploit the invention in Costa Rica or any other Central American country within three years from the date of issue;
- compulsory licensing in special cases (when no exploitation occurs; dependent patents; in case of public need);
- patent examinations are not made by the Registry but by professionals from professional associations and universities.

Proposed changes

Since Costa Rica became a WTO member in 1994, the Government became aware of the changes needed to comply with TRIPs requirements. The real questions became, does the law have to change? If so, when to make those changes? Since the Agreements on the Uruguay Round have become a law at the level of an international treaty, and the Costa Rican Constitution recognizes international agreements as ranking higher than national laws, debates have begun about whether TRIPs implementation needs changes in the law or if the laws were automatically changed when the National Assembly approved the Uruguay Round. For example, Costa Rican Patent Law establishes a 12 year protection period for a patent and the TRIPs agreement sets a 20 year protection period. If TRIPs is an international convention approved by the Congress, does that mean that the Costa Rican Registry, since that approval, has to automatically grant patents for a 20 year period? Or the Congress has to amend the Patent Law? There is still widespread disagreement in this matter but the thesis from the Government, saying that TRIPs implementation requires law modifications by the Congress, is prevailing.

The first big challenge for the Costa Rican IP system is to amend laws in accordance to TRIPs requirements. In terms of patents, this means basically allowing patentable subject matter to include microorganisms and biological processes with protection for animals and plant varieties. Also included are microbiological processes and their products. Compulsory licenses will also need regulation as required by TRIPs. But the most relevant change, in terms of negative impact on the national pharmaceutical industry, is amending the patent protection from a 12 year period to a 20 year period. This will have a substantial impact in pharmaceutical, agrochemical, fertilizer and food and beverage industries.

It is not clear when the changes on the Patent Law are going to be discussed. Two reform projects have so far been presented in Congress and are awaiting an agenda. Both projects try to fulfil TRIPs requirements.

As well as facing the challenge of implementing TRIPs, the Central American region and all hemispheric countries will have to address IPR issues in negotiations to create the Free Trade of the Americas region by the year 2005. Another area where IPR could be important is in the possible bilateral trade negotiations between the USA and the Central American countries.

Relationship with agriculture

Traditionally, agriculture has been an important sector in Costa Rica. The Costa Rican economy is based on two main export products: coffee and bananas. The coffee farms are mostly medium-sized and small, in contrast with banana production, which is owned by transnational companies with large fields. Agriculture is a fundamental area in the social and economic development of the country. The authorities are reluctant to protect these industries with patents, products and processes related to agriculture due to the lack of study on the impact this can have in this sector, especially price increases. Currently, agricultural goods such as seeds are not subject to protection in Costa Rica and this situation is not likely to change until the end of the transitional period given by TRIPs, which is the year 2000.

Costa Rica has a large number of researchers working on improvement of plant varieties, including transgenic materials. The country's efforts in this field are well recognized in Latin America, due to the high level of human resources and research facilities (Sittenfeld and Salazar, 1996). Moreover, many of these researchers were first to point out the consequences of not protecting the products of their research activities.

Costa Rican farmers use imported seed for some crops and domestic seeds for others. For coffee production, the country grows its own seed and there is a public research system that provides growers with wide access to new varieties and technologies. Because banana production is in the hands of transnational companies, these companies have their own research system which transfers technologies to all their small, commercial producers.

There are a number of small, successful companies involved in tissue culture and micropropagation. Costa Rica has the potential to benefit from biotechnology, especially from agricultural biotechnology. In addition, Costa Rica is a privileged country in terms of biodiversity and genetic resources, the raw materials for biotechnology (Salazar, 1992).

These are the facts that have to be taken into account when making decisions regarding IPR protection in agriculture. When dealing with the question about protecting agricultural biotechnology in developing countries like Costa Rica, some concerns arise. Developing countries want to assure their access to technology. Technology transfer is a key issue (Jorda, 1995). Developing countries are aware that innovation is crucial for development and that protection of IP is a basic step. However, they also do not want their farmers to pay high prices and limit their access to agricultural goods. So the decision is a difficult one (Salazar, 1995).

Another issue that has to be raised is the one related to the recognition of farmer's rights for the contribution they have made through the centuries improving varieties for the sake of human kind. An international debate has been taking place since the 1980s and no solution is yet proposed.

Plant variety protection

Plants, animals and biological processes are not patentable in Costa Rica, but there is a regulation in the Seeds Law that establishes that the Seeds Office has the obligation to create a protected variety registry and establish procedures that control plant breeders' rights. It seems that, when drafting this law, Costa Rica collaborated with Spanish consultants who influenced the regulation. However, at this moment there is no plant variety protection in Costa Rica.

Over the last two years, the Seeds Office has been working on the possibility of creating regulations needed to protect plant varieties using a system in accordance with the International Union for the Protection of New Varieties of Plants (UPOV). At this moment the office has a draft they plan to submit to Congress in the near future, in order to implement UPOV's 1978 Act.[1] A political decision on this matter has not yet been made, but all sides agree that in order to fulfil the TRIPs requirements of protecting plant varieties by patents, using a *sui generis* system, or both, Costa Rica must comply with UPOV.

LICENSING

Article 30 of the Central American Convention makes compulsory the registration of any sale, grant or license of a trademark. If everything is in order and the registration is subsequently published in the official

---

[1] There are differences between the 1978 and 1991 UPOV Acts. While there is a belief that the 1978 Act is more favorable to developing countries, countries only have a limited time in which to comply with this Act; if they fail to do so, they will have to comply with the later Act.

journal, the Registrar will dictate a resolution with a special note in the margin.

Table 13.2 indicates the number of trademarks and patents issued in a period of four years (1991–1994). The table shows the small proportion of patents granted, a situation which influences licensing activities. This situation has not changed much in the last two years.

According to Costa Rican Patent Law, licensing to third parties is the second most important right. As in trademarks, patent licenses have to be registered to be valid. Although the law does make compulsory licensing possible, no compulsory licenses have been issued on grounds of public utility.

CASE STUDY

Costa Rica is rich in biodiversity. This gives the country a special opportunity in terms of development. Traditionally, biodiversity was considered a natural resource, the heritage of the people. Ironically, with the development of biotechnology and the possibility of protecting biotechnological inventions with exclusive rights, large differences between developed and developing countries arose because biodiversity is considered a raw material for the development of biotechnological products. It is well known that geographic distribution of biodiversity is very uneven, with underdeveloped countries generally having the greatest diversity. The products derived from developing countries' biodiversity are transformed and patented in industrialized countries. The goods may have high commercial value and are sold and distributed without any compensation being paid to the country of origin. Following lengthy debate, most countries have now subscribed to the Biodiversity Convention (Asebey, 1996), which establishes the sole sovereignty of each state over its own biodiversity. Inspired by this idea, there is a

Table 13.2. The number of patents and trademarks applied for and issued in 1991–1994.

Category	Year	No. presented	No. issued	% issued
Patents	1991	186	35	18.8
	1992	134	24	17.9
	1993	129	10	7.8
	1994	181	53	29.3
Trademarks	1991	6269	3922	62.6
	1992	7021	3759	53.5
	1993	7598	3588	47.2
	1994	7562	3775	49.9

Wildlife Conservation Law in Costa Rica that establishes that biodiversity is in the public domain and of public interest. All wildlife is part of the national wealth, and any exploitation of the national biodiversity, such as extraction, production, commercialization, industrialization and use of genetic materials, is subject to the Ministry of Environment's authorization.

To improve the biodiversity-related legislation, the Costa Rican Congress is discussing two quite different projects of Biodiversity Law. The first intends to regulate biodiversity by restraining its access, while the second intends a more moderate approach. Academic and private sector participants are involved in these discussions, which are ongoing. Regardless of the level of regulation, if this biodiversity law is enacted in the near future, Costa Rica will be one of the first countries in the world to have a law of this type and to implement the possibilities achieved by developing countries in the Biodiversity Convention.

REFERENCES

Asebey, E. (1996) [Andes Pharmaceuticals, Inc.: a new model for biodiversity prospecting. In: *Biodiversity, Biotechnology and Sustainable Development in Health and Agriculture: Emergent Connections.*] World Health Organization, Washington, DC, USA, pp. 50–81 (in Spanish).

Jorda, K. (1995) [*Intellectual Property Rights: Reflections about its Importance.*] Paper presented at the Intellectual Property Seminar, San José, Costa Rica, 171 pp. (in Spanish).

Palacios, M.A. and Salazar, S. (1995) [*Diagnosis of Situation of Intellectual Property in Central America.*] SIECA, Guatemala, 502 pp. (in Spanish).

Salazar, S. (1992) *Plant Varieties and IPR, Regional Initiatives: a Developing Country's Perspective.* Paper presented at the Seminar on Plant Variety Protection, Patents and the GATT, the Philippines, 15 pp.

Salazar, S. (1995) [*Intellectual Property and Biotechnology.*] SIECA, Guatemala, 30 pp. (in Spanish).

Salazar, S. (1996) [*Intellectual Property and Access to Biodiversity.*] SIECA, Guatemala, 22 pp. (in Spanish).

SELA (1988) [*Regulations and Tendencies on Industrial Property Regimes and Policies.*] SELA, 9 pp. (in Spanish).

Sherwood, R. (1990) *Intellectual Property and Economic Development.* Westview Press, Inc., Boulder, Colorado, 226 pp.

Sittenfeld, A. and Salazar, S. (1996) [*Actual Situation of Agricultural Biotechnology in Costa Rica.*] Paper presented at the Regional Seminar on Planning, Priorities and Policies on Agricultural Biotechnology, Peru, 31 pp. (in Spanish).

Mexico **14**

José Luis Solleiro and Rosario Castañón

Center for Technological Innovation, National University of Mexico, Ciudad Universitaria, PO Box 22510, Mexico

INTRODUCTION

In the middle of this century, after significant growth in industrial property legislation, economists began to be concerned about the effects of the patents system and carried out the first studies on the subject. Developing countries, too, questioned the relevance of patent concession, above all in areas of technology considered to be strategic or of special importance for social welfare. In concrete terms, the capacity of the patent system to promote industrialization in these countries was put in doubt when it was seen that, except in some exceptional cases, patents were not exploited in developing countries, and were used defensively with the sole objective of maintaining monopolistic conditions in a market. There was also great concern over the effects of patent protection on prices. Several studies found that, under the temporary monopolistic protection of the patent, various firms marketed their products at prices far higher than the international ones.

In response to this situation, there arose in many developing countries a movement that sought to impose a defensive regime. Mexico, for example, in 1976, substituted its Industrial Property Law of 1942 for the Inventions and Trademarks Law. This Law excluded the possibility of granting patents in areas such as: (i) chemical products, all types of agrochemical, pharmochemical and pharmaceutical products, and the processes used to obtain them; (iii) technologies related to tackling pollution; (iii) food for human and animal consumption and the procedures used to obtain them. Similarly, with the intention of having a direct influence on the use of the technology patented in the country, industrial exploitation of the patented invention was made obligatory, with the risk of losing patent rights if this was not done within a period of three years from the date on which the patent was awarded. The period during which the patent would remain in force was also reduced to ten years from the date on which it was awarded and an aggressive compulsory licensing regime was introduced, together with something found nowhere else in the world, the invention certificate, which was granted in some of the areas excluded from patentability. As the invention certificate did not confer exclusive rights, it came under a compulsory licensing regime from the very outset.

While many developing countries adopted this defensive position, most industrialized countries, motivated by the constant emergence of new technologies and their growing importance, worked on the consolidation of an international system. Within this framework, great efforts were devoted to the establishment of agencies and international agreements such as the World Intellectual Property Organization, the Patent Cooperation Treaty, and the European Patent Office. The positions taken by the industrialized countries gave rise, for the first time, to the inclusion of a specific chapter on intellectual property (IP) in the Uruguay Round negotiations of the General Agreement on Tariffs and Trade (GATT). Taking the American bill as a basis, the industrialized countries began a search for an effective, sufficient level of protection for the intangible elements making up the value of a piece of merchandise. Lack of protection for IP would, in the US opinion, represent 'a significant, growing non-customs barrier for the trade in goods and services' (Correa, 1989). After drawn-out negotiations, the Trade-related Aspects of Intellectual Property Rights (TRIPs) Agreement was adopted in 1994 establishing minimum standards for IP protection in the member countries of GATT (now the World Trade Organization, or WTO). Nations that do not respect the levels of protection agreed upon will be the object of proceedings and, eventually, trade sanctions in other areas.

TRIPs is, at present, the most important international instrument for harmonization of IP legislation. Countries are now obliged to adopt minimum standards, and the flexibility and autonomy for defining national laws have been considerably reduced (Solleiro, 1997).

At the end of the 1980s, Mexico began negotiations to establish a North American Free Trade Agreement (NAFTA). One of the prerequisites laid down by the USA for the advancement of these negotiations was that Mexico changed its defensive IPR, copyrights, technology transfer and foreign investment laws. Under such pressure, Mexico changed the four laws.

RECENT CHANGES AND CURRENT STATUS OF MEXICO'S IPR LAWS

Trade liberalization policies were introduced in Mexico in the mid-1980s. At the same time, a heated debate also began on the possibility of reforming the Inventions and Trademarks Law of 1976. As result, the Law was reformed on 16 January 1987. The reforms introduced the possibility of granting patents to inventions for the protection of the environment, animal food, new processes to obtain alloys and processes to produce agrochemicals, pharmaceutical products and chemical products. The reforms also stated that, in a ten year period, patenting chemical products and almost any type of biotechnology invention would be allowed. The term of validity of patents was also extended from ten to 14 years.

Just a few years later, though, the Mexican government, as a response to the requirements for negotiating NAFTA, introduced profound changes in the IPR Laws. In June 1991, a new Law for the Promotion and Protection of Industrial Property was approved. This Law provides the possibility of granting patents to most inventions, according to standards established by TRIPs, and, at the same time, liberalized technology transfer by abrogating the Law for the Register and Control of Technology Transfer and the Use and Exploitation of Patents and Trademarks that had been in force since 1972.

In August 1994, with the objective of fulfilling the standards of the Intellectual Property Chapter of NAFTA as well as those of TRIPs, the Law for the Promotion and Protection of Industrial Property was amended. Its name was also changed to the Industrial Property Law. The reforms came into force in October 1994.

Main characteristics of the industrial property law

The following are the main provisions of the Industrial Property Law, at present in force.

1. As regards the industrial property rights contained in the 1994 Law, elements are established for (Article 1): (i) the award of patents, utility models and industrial design certificates for the protection of inventions; and (ii) trademarks, collective trademarks, notices and business names and appellation of origin for distinguishing signs of enterprises. Similarly, mechanisms are proposed for the repression of unfair competition by means of the clear, precise recognition of offenses and infringements; aspects related to the protection of industrial secrets are ratified, and mechanisms and criteria used in obtaining this protection are indicated, clearly specifying the type of information that is subject to these rights. It also includes aspects assisting technology transfer and the licensing of industrial property rights.

2. Inventions subject to patent protection are taken to be those that are new, the result of an inventive activity and that have an industrial application, with the exception of (Article 16): (i) essentially biological processes for the production, reproduction and propagation of plants and animals; (ii) biological material as found in nature; breeds of animals; the human body and the live parts it is composed of; and (iii) plant varieties.

3. Aspects not considered to be an invention for the purpose of the Law, are, in general terms, those commonly recognized by international practice, such as theoretical and scientific principles; discoveries; mental games and schemes, plans and commercial rules; computer programs; ways of presenting information and methods of surgical and therapeutic treatment and diagnosis to be applied to the human body or animals (Article 19).

4. Holders of patents and utility models are granted the right of prerogative to prevent other persons manufacturing, using, selling, putting up for sale or importing the patented product without their consent, and to prevent other persons using the process and using, selling, or putting up for sale or importing a product directly obtained from a patented process without their consent (Article 9).

5. Patents are valid for 20 years from the date of application (Article 23).

6. Utility models include inventions related to utensils, apparatus or tools whose layout, configuration, structure or shape has been modified and that have a different function with respect to the parts from which they are made or advantages as to use (Article 28).

7. Utility models are valid for ten years from the date on which application was made (Article 29).

8. With respect to industrial design registers, two- and three-dimensional shapes of no determined technical function can be protected. These registers are valid for 15 years as of the date of application (Articles 32 and 36).

9. In the case of trademarks, registers are awarded for names and two- and three-dimensional logotypes, as with collective trademarks. These registers are valid for ten years from the date on which application was made and can be renewed for successive ten-year periods (Articles 89 and 95).

10. An industrial secret is considered to be all information, of any type, that allows an enterprise to obtain and/or maintain competitive advantage over third parties. Necessary measures need to be adopted to conserve confidentiality of the said information and it also needs to be fixed on to a material support[1] (Article 82).

---

[1] Information under industrial secret must be registered on paper, electronic or magnetic format.

11. As in the legislation of other countries, privileges are also granted for commercial slogans, tradenames and appellations of origin (Fourth Document, Chapters 3 and 4; Fifth Document).

12. Sanctions for industrial property offenses include imprisonment (from two to six years) and fines amounting to the equivalent of one hundred to 10,000 times the daily minimum wage. In addition, redress and payment for damages may be claimed, and shall, in no case, be less than 40% of sale price to the public (Articles 224, 226 and 221*bis*).

There are multiple differences between this Law and preceding ones. The most important of these include the validity of the different instruments, the possibility of patenting in various areas and, above all, greater stringency in penalizing offenses in this matter.

Plant variety protection law

As mentioned above, in accordance with TRIPs and NAFTA, Mexico decided in 1994 to take the *sui generis* option to protect new plant varieties. After a long discussion with different stakeholders (seed industries, plant breeders, producers associations, etc.) the Congress of the Union approved the Federal Plant Variety Law, published in the official *Gazette* on 25 October 1996. In general, the Law corresponds to the provisions of UPOV 1978 (Mexican Congress ratified the country's adhesion to UPOV 1978 in December 1995). The Ministry of Agriculture, Livestock and Rural Development (*Secretaría de Agricultura, Ganadería y Desarrollo Rural*) is responsible for administering it. The main provisions of the Law are the following:

1. The rights granted under this law to plant variety breeders are (Article 4) to be recognized as a plant variety breeder; and to avail themselves of and exploit, on their own or by third parties with their consent, a plant variety and its propagating material, for the reproduction, distribution or sale thereof, as well as for the production of other plant varieties and hybrids for commercial purposes. These rights will have a term of 18 years for perennial species (forest trees, fruit plants, vines, ornamentals) and their rootstocks, and 15 years for those species which are not included in the former category.

2. The consent of a plant variety breeder is not required to use the same (Article 5) as a source or raw material for research related to the genetic improvement of other plant varieties, in the multiplication of propagating material for self-use as grain for consumption or sowing, or for human or animal consumption, exclusively to the benefit of the person harvesting the same.

3. The requisites for obtaining the breeder certificate establish that the variety must be (Article 7): (i) 'new', i.e. must not previously have been put up for sale or marketed; (ii) 'distinct', i.e. having one or several pertinent characteristics that technically and clearly distinguish it from any other variety; (iii) 'stable', i.e. having pertinent characteristics that remain unchanged after successive reproductions and propagations; and (iv) homogeneous.

4. Article 9 states that, in an application for a breeder certificate, a name must be proposed for the variety.

5. Priority shall be granted to the applicant of a breeder certificate who has previously filed the same application in the foreign country with which Mexico has had or will have agreements or treaties in this field (Article 10). The priority shall consist of possible recognition as filing date, the date when the application was filed in another country, provided 12 months have not expired.

6. In order to recognize the priority referred to in the preceding Article, the following requirements shall be complied with (Article 11):

- that at the time of applying for the breeder certificate, priority is claimed, and the country of origin and filing date of the application in that country are mentioned;
- that the application filed in Mexico is not with the purpose of obtaining rights to those derived from the application filed abroad; and
- that within three months from the filing date of the application, the requirements set forth in international treaties, this law and its regulations are complied with.

7. In Article 25, the Law establishes particular provisions for granting emergency licenses. Thus, it is understood that there are emergency circumstances, when the exploitation of a plant variety is considered to be essential to satisfy the basic needs of one sector of the population and there are deficiencies in its demand or supply. If a plant variety is not exploited within three years from the date of issue of the breeder certificate, this case will be considered as an emergency.

8. The competent body for the administration of the Law is the Ministry of Agriculture (SAGAR) through the National Service for Seed Inspection and Certification (SNICS). According to the Third Document of the Law, SAGAR will be assisted by the Plant Variety Assessment Committee whose main duties relate to (Article 30): examination of the merits of applications for breeder certificate and their registration; to establish the procedures to carry out and evaluate technical field or laboratory tests; to render an opinion for the elaboration of Official Mexican Standards, related to the characterization and evaluation of plant varieties for description purposes; and all other duties set forth in the bylaws.

9. SAGAR will establish a public register, in which mainly the following items shall be recorded: applications, filing certificates and breeder certificates.

10. According to Article 48, SAGAR shall impose the following fines, in accordance with the Federal Law of Administrative Procedures for the infringements set out below:

- for modifying the denomination of the protected plant variety, without the Department's authorization, an amount equivalent to 200–2000 times the minimum daily wage;
- for misrepresenting itself as the holder of a protected plant variety, when it is not, an amount equivalent to 500–3000 times the minimum daily wage;
- for divulging or marketing a plant variety as of foreign origin when it is not, or for divulging or marketing a plant variety as of domestic origin, when it is not, an amount equivalent to 300–3000 times the minimum daily wage;
- for opposing the verification inspections made in accordance with this law and the Federal Law of Administrative Procedure, from 300 to 3000 times the minimum daily wage;
- for commercially exploiting the characteristics or content of a protected plant variety, attributing them to an unprotected plant variety, from 1000 to 10,000 times the minimum daily wage;
- for ceasing to comply with or for violating the measures established in Article 42 of this law, from 1000 to 10,000 times the minimum daily wage;
- for taking advantage of or exploiting a protected plant variety, or its propagating material, for production, distribution or sale without the holder's authorization, from 2000 to 10,000 times the minimum daily wage; and
- for all other infringements to the provisions of this law and its regulations, from 200 to 5000 times the minimum daily wage.

Federal Copyright Law[2]

Over the last few years, above all as a result of signing NAFTA, there has been great pressure on Mexico to modify the law, since it was considered that an appropriate legal framework for the protection of works that were the product of the intellect did not exist (among other things, greater protection in the area of computation, in particular the creation of computer programs and databases, was desired). The new Federal Copyright Law was negotiated for more than two years and after various modifications was unanimously passed on 24 December 1996. It is important to point out that this new law designates a new administrative

---

[2] A right related to the economic benefits deriving from commercial exploitation of intellectual property rights.

body for its application, namely the National Institute for Copyrights, which has not yet been created. The Mexican Industrial Property Institute will also intervene, particularly in matters relating to sanctions for infringement of a commercial nature. One of the innovations of this Law was the corresponding reform of the penal code in order to typify offenses in this respect and to increase penalties for persons that commit the said offenses, especially those related to piracy (Castro *et al.*, 1996).

Matters relating to copyrights are regulated by the Public Education Ministry. The most important points of the new Law are given below.

1. In order for a work to be protected by this Law, it needs to be fixed on to a material support (Article 5).

2. National and foreign authors (or rights holders) are entitled to the same rights. Similarly, international treaties on this matter are recognized (Article 7).

3. Authors' rights include works in the following fields: literature, music (with or without words), drama, dance, pictures or drawings, sculpture, cartoons and cartoon stories, architecture, cinematography, graphic and textile design, collections (of works such as encyclopedias, anthologies, databases, etc.) (Article 13).

4. The following are not subject to protection as copyright: (i) ideas in themselves, formulae, concepts, methods, systems, discoveries, processes and inventions; (ii) industrial or commercial use of ideas contained in the works; (iii) games; (iv) isolated words, digits or colors; (v) isolated names, titles or sentences; (vi) blank forms with instructions; (vii) reproductions of emblems from any country, state, municipality or equivalent political division; (viii) legislative texts; (ix) news (but its means of expression is included); (x) information commonly used, such as sayings, legends and metric scales (Article 14).

5. The Law recognizes the moral and patrimonial rights of the author (Article 18 and 24).

6. The holders may authorize or prohibit: (i) the reproduction, publication or edition of the work; (ii) public communication of their work; (iii) public transmission or broadcast of their work; (iv) distribution of the work; (v) publication of derived works (Articles 26 and 27).

7. Patrimonial rights shall be valid during: (i) the life of the author and 75 years from the time of his/her death; (ii) 75 years after publication of posthumous works as well as work done as part of an official service for the Federation (Article 29).

8. Computer programs are protected under the same terms as literary works. The holder of copyrights to a computer program shall keep, even after the sale of copies of the same, the right to authorize or prohibit the leasing of the said examples. This precept shall not apply when the copy of the computer program does not constitute an essential object of the license for use. Databases will be protected as compilations (Articles 102, 104 and 107).

9. Penalties in case of infringement of copyrights can be fines ranging from the equivalent of 1000 to 15,000 times the minimum daily wage, while penalties in case of infringements of a commercial nature can be fines ranging from 500 to 10,000 times the minimum daily wage (Articles 230 and 232).

Foreign Investment Law

This Law was passed on 27 December 1993, to replace the Law to Promote Mexican Investment and Regulate Foreign Investment, which had been passed in 1973. The Ministry empowered to supervise compliance with the provisions contained in this Law is the Ministry of Trade and Industrial Development (SECOFI). There can be no doubt that this Law was one of the first to be modified in order to give consistency and legitimacy to the economic model introduced during the last government administration – that of an unrestricted free market. Similarly, the pressure imposed by other countries in the framework of the commercial agreement of the GATT on direct foreign investment and, in particular, by Mexico's trading partners in NAFTA, can clearly be seen from the modifications introduced. The most important points of this Law are given below:

1. There is no limit to foreign participation in the social capital of Mexican companies with respect to their purchase of fixed assets, manufacture of new products, or the broadening of already existing lines (Article 4).

2. Foreign participation is excluded from the following economic activities: oil and basic petrochemicals; electricity; generation of nuclear energy; radioactive minerals; satellite communication; telegraphs; mail; railways; issuing of bank notes and coining of money; control, supervision and monitoring of ports, airports and heliports; national ground transport (excluding courier services); retail trade of gasoline and liquid pressure (LP) gas; broadcasting services other than cable television; credit unions; and development bank institutions (Articles 5 and 6). Recent measures have gradually opened the possibility of investing in these reserved sectors.

3. Economic activities in which foreign participation is limited (percentages vary) are: cooperative production organizations, national air transport, multibank credit institutions, stock exchanges, bonding institutions, exchange bureaus, financial leasing companies, financial factoring firms, manufacture and commercialization of explosives and related products, printing and publication of newspapers for exclusive circulation in national territory, shares in companies that own agricultural, livestock and forest land, cable television, basic telephony services, fishing, shipping companies, and services related to the railway sector (Article 7). In some cases, it is possible to request a greater participation of foreign investment. This decision must be taken by the National Commission for Foreign Investment.

4. The constitution of companies and changes in them must be approved by the Foreign Relations Ministry (Articles 15 and 16).

IPR laws and promotion of innovative activities in Mexico

The justification normally given for introducing and strengthening IP legislation in countries with a lower level of industrial and technological development is that, without the incentive of market protection these rights represent, the flow of investment, trade and technology towards these countries would be interrupted. It is also argued that lack of effective protection would inhibit the innovative capacity of society. On the other hand, opponents of the industrial property system state that monopolistic protection is exclusively used to reserve exclusive import markets, without investments and efforts to develop productive investment and innovation in the country in question.

It is true that there are not sufficient studies to make it possible to come to conclusions on these hypotheses. Furthermore, the problem of assessing the effects of IP protection on the economy and society is very complex, since it is virtually impossible to isolate the phenomenon as an object of study, and separate it from the extremely broad context of economic, sectorial, industrial, agricultural and science and technology policy. However, it is possible to see that the effects and strategic importance of IP differ considerably from sector to sector. In a recent study, Mansfield (1992) revealed that firms that consider IPR more important for their investment decisions in new manufacturing installations tend to be large and intensive in research and development.

It can in no way be assumed that the simple introduction of modern legislation governing IP will be a sufficient condition to attract foreign (or national) capital to a certain sector. At present, there seems to be an agreement that, due to the current conditions in international trade, this is a requirement, but other, perhaps more important, factors are needed to create the competitive environment necessary to attract investment, such as rapid economic growth, low relative costs, political and social stability, supply of special capacities in some sector of the economy, and the existence of political advantages and specific development program.

There can be no doubt that Mexico has adapted its legal framework to present international demands that are determined by market globalization and high technology dependence. Similarly, great efforts have been made to create the necessary bodies to execute the legal provisions: the Mexican Industrial Property Institute, the National Authors' Rights Institute, the National Service for Seed Inspection and Certification, and the National Commission for Foreign Investment. Nevertheless, for the modern IP framework to yield fruit, the country's innovating capacity must be substantially increased in the short term, otherwise the most probable situation is that IP titles will be used basically to import and distribute the new technology products in an exclusive way. The technologies themselves will certainly be disseminated but principally towards sectors that are highly profitable and economically attractive. The stark reality, especially in the agriculture sector, is that there is a very large group of producers with low incomes and few technological resources for whom the news is not so good. The protection of generic technologies and even of research tools by patents and industrial secrets may be a considerable barrier for gaining access to the new technologies. In synthesis, effective protection will facilitate access to technology for those who are already in an advantageous position. The same cannot be expected for those who do not have the same resources; for them it will be an entry barrier instead.

REFERENCES

Castro, J., Terrazas, A. and De la Vega, M. (1996) Obligada por el TLC, la visión mercantilista borró el humanismo en la nueva ley autoral. *Proceso* 1048, 52–57.

Correa, C.M. (1989) Propiedad intelectual, innovación tecnológica y comercio internacional. *Comercio Exterior* 39(12), 1059–1082.

Mansfield, E. (1992) Unauthorized use of intellectual property: effects on investment, technology transfer and innovation. Cited by Correa, C.M. (1995) Intellectual property rights and foreign direct investment. *International Journal of Technology Management* 10(2/3), 173–199.

Solleiro, J.L. (1997) Intellectual property rights and the growth of biotechnology based industries in developing countries. *Biotechnology Advances* 15(3/4), 565–582.

European Union

R. Stephen Crespi

*Patent Consultant, 16 Kenlegh, Bognor Regis
PO21 3TS, UK*

INTRODUCTION

For more than a decade it has been a matter of dispute whether plants can be the subject of patent protection, in addition to or as an alternative to the protection afforded by plant variety rights. This was one of many questions in patent law to which no single global answer could be given, owing to the differences of law from one country to another. Under the laws of the USA and Australia, for example, a clear affirmative statement can be made, subject of course to meeting the basic conditions of patentability which apply to any invention. But in Europe and most other countries it has been more difficult to answer this question clearly and simply.

THE EUROPEAN AND INTERNATIONAL LEGAL BACKGROUND

Plant and animal varieties

In Europe the patent law was originally considered unsuitable for protecting new plant varieties developed by traditional breeding methods. Special national laws of plant breeders' rights (also called plant variety rights) were therefore established in the 1960s in some countries and an international convention, the International Union for the Protection of New Varieties of Plant (UPOV, 1961), was formed.

Because plant breeders' rights were a major innovation, and to an extent controversial in agricultural circles, they were consciously designed to provide a form of protection less strong than that of patents. For example, two freedoms were enshrined in the law, one express, the other implied. First, it was expressly stated that breeders were free to use a legally protected variety as a starting point for breeding further varieties, i.e. they could do so without payment of a royalty. This was known as the 'breeder's privilege' or 'research exemption'. Secondly, because the rights were restricted to commercial dealing in the reproductive material of the variety, a farmer sowing purchased seed of the variety was implicitly free to save seed from the harvest for subsequent sowing on his own farm. This was the 'farmer's privilege'.

Plant breeders' rights have been highly successful in their own sphere. However, it is now generally recognized that patent law is the better suited to the protection of recombinant methods for producing transgenic plants and the resulting products. Patents of this type, claiming methods and products *per se*, have been granted by the European Patent Office (EPO). Unfortunately, recent EPO case law has introduced some uncertainty into the legal situation which may take time to resolve.

Animal breeds produced by traditional methods have no legal system for their protection comparable to plant breeders' rights. Following the declaration by the US Commissioner of Patents in 1987 that US patents would be granted for 'non-naturally occurring non-human multicellular living organisms including animals', the first transgenic animal patent issued in 1988 to Harvard University with claims covering the 'onco-mouse'. After initial reluctance by the EPO to grant the corresponding European patent (and a successful appeal to the Appeal Board) the European patent was issued. This is now under formal opposition by antivivisection and animal rights groups.

European patent laws

All European countries have their own national patent law and most are also members of the regional patent system of the European Patent Convention (EPC, 1973). Under the EPC, a single patent application can cover all, or any selection, of the countries that have joined this Convention. EPC law takes precedence over national laws and these are required to be in harmony with it.

In addition there is the politico-economic grouping of the European Community or Union (EU) which can legislate for EU members by Directives or Regulations. Examples are the proposed EU Directive on the Legal Protection of Biotechnological Inventions (EU Directive, 1988/1995), and the European Council Regulation on a Communitywide system of plant variety rights (European Council, 1994). Most European countries have national laws of plant variety protection and are also members of UPOV. UPOV has been revised more than once since its inception. The currently operative text is the 1978 version (UPOV, 1978). A further significant revision was made in 1991 (UPOV, 1991) and awaits ratification by member states. This complex mix of applicable laws gives rise to the legal interface problem.

The interface between patent and variety protection

This question has been addressed jointly by the World Intellectual Property Organization (WIPO) and by UPOV in order to determine whether the patent system and the plant variety right system are incompatible or complementary, each operating in a defined sphere (WIPO/UPOV, 1990). The question is important to patent law owing to the exclusion of plant and animal varieties from patent protection in some countries. For example, Article 53(b) of the EPC prohibits patents for 'plant or animal varieties or essentially biological processes for the production of plants or animals: this provision does not apply to microbiological processes or the products thereof'. National patent laws in European countries contain the same provision.

It is noteworthy that the second half of Article 53(b) limits the exclusion. It is believed that this was included to safeguard the patentability of microbial cultivation methods and resulting products, e.g. antibiotics.

The term 'essentially biological' has not yet been judicially defined although, as mentioned later, some attempt at clarification has been made in the EPC case law. Bearing in mind the birth of the UPOV legislation, and the desire to ensure that patents would not impinge on plant breeding methods, the term may have been simply intended to apply to the traditional processes used to breed new plant varieties. In spite of the confusion to which this term has given rise, the legislators seem unable or unwilling to dispense with it.

What is a plant variety?
First there is a problem of semantics. To the plant scientist the term 'variety' is not a botanical taxon and lacks scientific precision. To plant breeders, the term 'variety' served well for practical purposes and was apparently used rather flexibly, without the need for a rigid definition. The Vice Secretary-General of UPOV has stated that

> The variety was an abstract concept which had been developed by users of plant varieties such as agriculturalists and researchers such as botanists and taxonomists to assist in the classification of plant material. The concept was not a concise one. It had no existence on its own. ... Many rules had been established to define the unit of plant material that would be considered as

a variety, mainly in terms of the mechanism used for reproduction or propagation.

(WIPO/UPOV, 1990)

The definition of the plant variety, used in the original 1961 version of the UPOV Convention, in Article 2(2) stated that 'For the purposes of this Convention, the word "variety" applies to any cultivar, clone, line, stock or hybrid which is capable of cultivation and which satisfies the provisions of sub-paragraphs 1(c) and 1(d) of Article 2'. (The cited sub-paragraphs dealt with homogeneity and stability.) According to this definition, then, a variety was whatever satisfied the criteria of distinctness, uniformity and stability and was therefore protectable under the UPOV Convention. This definition was removed when the Convention was revised in 1978.

The above arrangement seemed to work satisfactorily for almost two decades. The UPOV system was protected from any interfacial tension with the patent system by its own prohibition of protection by both forms ('double protection') in Article 2(1), which provided

> Each member State of the Union may recognise the right of the breeder provided for in this Convention by the grant either of a special title of protection or of a patent. Nevertheless, a member State of the Union whose national law admits of protection under both these forms may provide only one of them for one and the same botanical species or genus.

This restriction was reinforced in the patent laws of those countries that had expressly excluded plant varieties from patent protection, e.g. according to the prototype provision of EPC Article 53(b). Since the EPC came into being, very few attempts have been made to disturb the situation by filing patent applications for plant varieties as such.

The exclusion of plant varieties from patentability

In Europe the area excluded from patent protection was identified as co-terminous with the area covered by the UPOV system at the time the point arose for decision in the Ciba-Geigy case (Ciba-Geigy, 1984). The claim before the EPO related to 'Propagating material for cultivated plants, treated with an oxime derivative according to (a specified) formula ...'.

In allowing this claim, the Technical Board of Appeals held that 'Article 53(b) prohibited only the patenting of plants or their propagating material in the genetically fixed form of the plant variety'. Here the claimed propagating material was the result of treatment with a seed dressing agent to protect seeds against certain herbicides. It was not a variety and neither was it the result of an essentially biological process for the breeding of plants.

Two further passages in the Board's decision are especially noteworthy. Referring to the 1961 definition of a variety the Board observed that, 'The legislator did not wish to afford patent protection under the European Patent Convention to plant varieties of this kind, whether in the form of propagating material or of the plant itself'. With reference to the claim in dispute, the Board observed:

> It is immaterial to the question of patentability that the propagating material which is treated can also be, or is primarily, a plant variety. If plant varieties have been excluded from patent protection because specifically the achievement involved in breeding a new variety is to have its own form of protection, it is perfectly sufficient for the exclusion to be left restricted, in conformity with its wording, to cases in which plants are characterised precisely by the genetically determined peculiarities of their natural phenotype. In this respect there is no conflict between areas reserved for national protection of varieties and the field of application of the EPC. On the other hand, innovations which cannot be given the protection afforded to varieties are still patentable if the general prerequisites are met.

This approach to Article 53(b) was consolidated in the case of Lubrizol Genetics Inc. (Lubrizol, 1990). In the process claimed in this application, parent plants with desired characteristics are selected, test-crossed, marked and stored. The hybrids resulting from the crosses are then evaluated for desired traits and phenotypical uniformity and that pair of parent plants (at least one of which is heterozygous) which provides the desired hybrids is selected. At least the heterozygous parent plant is multiplied by cloning and the crossing of the said pair of parent plants is repeated as often as desired to provide hybrid plants on a large scale. The Technical Board of Appeal considered that, in a multi-step process, each single step as such may be characterized as biological in a scientific sense. However, in this case the essence of the claimed process lay in the particular combination of specific steps. The totality and the sequence of the specified operation neither occurred in nature nor corresponded to classical breeders' processes. The arrangement of steps in the claimed process represented an essential modification of known biological and classical breeders' processes, and the efficiency and high yield associated with the product showed important technological character.

The Board held that in Article 53(b) the exclusion of 'essentially biological' processes for the production of plants and animals should be construed narrowly. Whether or not a process is to be considered as 'essentially biological' has to be judged on the basis of the essence of the invention taking into account the totality of human intervention and its impact on the result achieved.

It was also decided that the products of this process could be claimed in 'product-by process' terms. Such products were not 'plant varieties' and therefore were not excluded as such under the first part of Article 53(b). The conclusion of the Board on this point was based on the fact that the hybrid seeds or plants produced by this process, though phenotypically uniform, would not breed true, i.e. did not possess the degree of stability necessary for them to be classed as varieties. This reasoning must have seemed rather puzzling to the applicant, who had no doubt that his strategy was in fact to produce 'hybrid varieties' although this term was not emphasized in the specification.

Thus, according to the European case law, especially the Ciba-Geigy case, the excluded area was to be equated with that which is protectable under UPOV and corresponding national laws of plant variety protection. Plants which have been specially bred as a new variety were to be protectable by plant breeders' rights if criteria of distinctness, uniformity and stability were met and they were *de facto* excluded from patent protection under patent laws in Europe. This conclusion was seen as conforming to the then-prevailing UPOV ban on double protection for the same entities.

It was therefore the understanding in patent circles that a variety was a subgroup of a plant species (or subspecies) containing individual members which resembled one another phenotypically and complied, for the most part, with a set of listed characteristics which constituted the official description of a protected variety by which it was distinguished from other such subgroups of the same species. Patent law could live comfortably with such a notion. With the advent of plant biotechnology, patent specialists argued that the above exclusions could not apply to recombinant DNA methods and transgenic plants.

The European Commission's proposed Directive

The European Commission's proposal in October 1988, for a Directive to EC Member States concerning the legal protection of biotechnological inventions accepted the patent case law outlined above as its starting point. In order to ensure that patent protection was available for inventions in plant biotechnology, Article 3 of the Commission's original text of the Directive provided that 'biological classifications other than plant or animal varieties ... shall be considered patentable subject matter'. In the course of some years of discussion with official representatives of Member States, this formulation had been modified. After receiving a negative opinion on the Directive by the European Parliament (October 1992) followed by a total Parliamentary rejection of the Directive (March 1995), the European Commission revamped many of the controversial Articles and resubmitted a new version of the Directive. The current draft (now Article 4(2)) reads: 'Biological material, including plants and animals, as well as elements of plants and animals obtained by means of a process not essentially biological, except plant and animal varieties as such, shall be patentable'.

In the original discussions with Member States it became necessary to dispose of a particular argument which has assumed significance. If a patent claim to a plant classification higher than the variety, e.g. a species, is to be construed as directed to the sum total of all conceivable varieties which would possess the characteristics defined in the claim, it was argued that a claim to a collection of individual varieties must, *a fortiori*, be excluded from patentability. It was argued by the present contributor that generic claims in patents are not directed to 'collections' in this fashion. Rather, such claims pinpoint the essence of the invention common to all the entities embraced thereby. Secondly, the subdivision of the claim into varieties was an arbitrary one. A more correct analysis would be to conceive a generic claim to, say, a transgenic plant as embracing all plants falling within its scope, irrespective of whether or not such plants belong to any particular variety.

PLANT VARIETY RIGHTS: RECENT DEVELOPMENTS

Two major developments took place after the above-mentioned patent case law was established, namely, the Revision of the UPOV Convention in 1991 (UPOV, 1991) and the European Commission's Regulation on Community Plant Variety Rights (European Council, 1994).

The European Community Plant Variety Right

The idea of a European Community Plant Variety Right stemmed from an initiative of the European Community Directorate responsible for agriculture. According to Article 1 of the Regulation, the system is to be established as 'the sole and exclusive form of Community industrial property rights for plant varieties'. A preceding explanatory recital ('Whereas ... etc.') states that the Regulation 'implements the ban on patenting plant varieties only to the extent that the European Patent Convention so requires, i.e. to plant varieties as such'.

The 1991 revision of UPOV

The protection given under UPOV has been improved and strengthened by this revision (Byrne, 1991). The prohibition of double protection in former Article 2 has been removed, although Member States retain the power to preserve this prohibition in their national laws.

A carefully worded definition of a plant variety now stands at the forefront of this Convention in Article 1 (vi). It states:

'variety' means a plant grouping within a single botanical taxon of the lowest known rank, which grouping, irrespective of whether the conditions for the grant of a breeder's right are fully met, can be

- defined by the expression of the characteristics resulting from a given genotype or combination of genotypes;
- distinguished from any other plant grouping by the expression of at least one of the said characteristics; and
- considered as a unit with regard to its suitability for being propagated unchanged.

UPOV spokesmen insisted that a definition conforming to the usage of the term in the agricultural industry was essential in present-day circumstances. The new definition is no longer to be equated with 'UPOV-protectable variety'.

Another respect in which protection under UPOV has been widened is that, under Article 14 (Scope of the Breeder's Right), the right is to extend to 'essentially derived varieties'. The complex definition of this term given in Article 14(5) will not be discussed here (see Byrne, 1991). However, the Vice-Secretary of UPOV has declared the view that it would cover a genetically modified variety which retains the whole genome of the original protected variety.

The freedom for breeders, farmers and for research

Freedoms for breeders and farmers are seen by some (mainly those opposed to intellectual property) as threatened by intellectual property systems, especially by patents on transgenic plants and animals.

The breeder's privilege

The 'breeder's privilege' or 'research exemption' noted above gave breeders the freedom not only to use protected plant varieties in their breeding programs but also to commercialize the further varieties developed therefrom (often only 'cosmetically' different from the original) without any royalty payment to the owner of the initial variety. This freedom is modified in UPOV 1991. The first part of this freedom has been retained in Article 15(1), which provides that the breeder's right does not extend to 'acts done for the purpose of breeding other varieties'. The second part has been curtailed as a result of the broadening of the scope of the right to 'essentially derived' varieties.

The freedom to research is safeguarded equally under both patent law and plant variety right (PVR) law, but the freedom to commercialize the resulting new varieties depends on whether or not they infringe the patent claims or are 'essentially derived varieties' under PVR law. The strengthened UPOV-type protection therefore goes part of the way towards the strong protection given by patents. Incidentally, neither system is a threat to the free use of existing germplasm since these rights can in no sense monopolize known material as such.

The farmer's privilege
The ability to save and re-sow seed, as explained above, was a consequence of the restricted definition of the scope of the breeder's right. Recognizing that the current scale of use of farm-saved seed thus deprives the breeder of significant royalty income, the strengthened right under Article 14 of UPOV 1991 now makes all propagation subject to the authorization of the breeder. However, Contracting States can 're-introduce' this freedom under their national legislation 'within reasonable limits and subject to the safeguarding of the legitimate interests of the breeder'. According to Article 14(3) of the European Community Plant Variety Right regulation the royalty rate on re-use of saved seed is to be 'sensibly lower' than that for bought-in seed. Until the UPOV 1991 revision is taken up in national laws in order to supersede UPOV 1978, though, farmers legitimately sowing seed of a protected variety still enjoy the 'farmer's privilege'.

Now that the 1991 UPOV no longer prohibits the availability of both types of legal right (patent and PVR) plant breeders who are themselves using the techniques of biotechnology alongside traditional breeding methods will wish to obtain both types of protection as appropriate.

Invention, protection and exploitation

The legal principles discussed above may be better appreciated in the light of a concrete practical example. This example is taken from European Patent Publication no. 272,144 (also US patent no. 5,306,863) but the claims presented below have been drafted by the author for the purposes of the present discussion.

The gene responsible for producing a trypsin inhibitor in the cowpea (*Vigna unguiculata*) has been transferred to other genera of plants. The cowpea is a legume, also called black-eyed bean, which is grown as a food crop in West Africa and in both North and South America. The trypsin inhibitor produced by resistant varieties of this plant prevents invading insects from digesting protein so that they die of starvation. Transfer of the inhibitor gene to other plant genera requires the methods of plant biotechnology and cannot be achieved by traditional breeding methods. The technology is aimed at protecting cotton and cereals against bollworms of the genera *Heliothis* and *Anthonomus* which affect these crops throughout the American and African continents. It is applicable also to protect grain of wheat, maize (corn), rice and sorghum against storage pests of the genera *Tribolium*, *Sitophilus* and *Chilo*, the latter being particularly serious in Africa, India, China and Japan.

Considering this invention first from the aspect of patenting transgenic cotton plants, the following types of claim are conceivable:

- a transgenic cotton plant having a gene for a trypsin inhibitor (type 1);
- a transgenic cotton plant having a gene for a trypsin inhibitor derived from the cowpea (type 2);.
- a cotton plant of the variety Stoneville 825 containing a gene for a trypsin inhibitor derived from the cowpea (type 3).

How should these claims be treated in official examination by patent authorities? Before the most recently decided EPO case law to be described below, the EPO would allow claims of type 1 and 2 because the plants are not claimed at the varietal level of definition. Each of these claims will cover all manner of varieties of cotton in which the gene has been introduced but patentability should not be affected by this fact. The claims should be allowable or not depending on whether or not they express an invention, and the plants covered by the claims are not in any sense being patented as varieties but as articles embodying an inventive step.

Claim type 3 above is the only claim which mentions a variety and is thereby arguably open to objection. It is a strange result that the patent applicant is apparently barred from specifically claiming the application of his invention to a particular commercially important variety. Since the major crop plants are marketed as varieties, what use would a transgenic plant patent be if it did not cover such an application? This anomalous result is one unforeseen consequence of the desire to draw an absolute line between the two forms of legal protection. The transgenetic process, whereby the foreign gene enters the genome of the starting variety, will not necessarily result in another variety in the older sense of the term, i.e. a distinct, uniform, stable variety. The process will produce the parental material from which further varieties will be bred. However, as a result of the new variety definition in UPOV 1991, the EPO have changed their attitude to patent claims of the above type.

Commercial exploitation

A typical pattern of the creation and exploitation of this type of technology could be as follows. A biotechnology research group in a scientific research institution or in an industrial research laboratory will have isolated the gene from the germplasm of the source country and will have patented the gene construct and the method of gene transfer to the plants targeted for protection. The patent owner will be free to develop and exploit this technology commercially on his own behalf. But it may be better to license the technology to commercial plant breeders in industrially developed countries and to appropriate organizations in developing countries, e.g. state-run agricultural research institutes, together with the know-how to transfer the gene to chosen types of plant. The plant breeders or research institutes may obtain plant breeders' rights for any resulting varieties. The new varieties will be sold to farmers who will cultivate them and, as a result of their improved pest resistance, will be able to economize in the use of chemical pesticides. The public will benefit from the advantages to the environment resulting from this technology. It is difficult to see who will not gain from this achievement. Unless the transgenic plant enables the farmer to achieve a better yield or a saving on the use of insecticides it will not be worth the higher price asked for it and it will not be purchased.

It remains to be seen whether the commercial exploitation of these forms of legal right, either as alternatives or in combination, can be managed successfully without undue burden on farmers and other end-users. In the plant field the negotiation of commercially reasonable royalty rates on farm-saved seed should not be unduly difficult and would avoid breeders having to ask high prices on the original sale of seed in order to recoup their investment in research and development in single payments. In the case of transgenic farm animals intended as breeding stock it would be less easy to enforce rights through successive generations and the animal breeders may well have to be innovative in devising commercially feasible methods of ensuring a return on their investment. In all cases, however, the continuing need to compete with traditional varieties and breeds ought to induce patent-holders and PVR-holders to follow reasonable policies. The 'Abuse of Monopoly' provisions which exist in both legal systems should also work towards the outcome of common sense.

SHOULD PLANT VARIETIES REMAIN UNPATENTABLE?

The view has been held for some years in industrial and patent professional circles that a plant variety should be patentable provided it meets the criteria of patent law. It has also been urged that both types of protection should be allowed (cumulative protection) provided the criteria under each system are fulfilled. These ideas began to gain a hearing in official patent circles, a noteworthy development which was encouraging to those who have held their ground throughout this time (Straus, 1984). It should be noted that the suggestion applies to the patenting of the specific variety, as such, and would therefore require the abolition of provisions such as EPC Article 53(b).

Realistic commentators admit that most varieties of the kind typically presented for plant variety protection will not qualify for patent protection because of the difficulty of showing that they entail an inventive step. It would also be difficult to describe the method of breeding in a way that would be repeatable. Therefore the PVR should remain as the preferred option for legal protection for innovations at the level of specific varieties.

The lack of examples of attempts to patent plant varieties of the typical kind for which plant variety rights are granted has tended to give this debate an academic rather than practical character. However, the rejection by the Supreme Court of Canada of a patent application for a soybean variety produced by methods of cross-breeding and selection (Pioneer Hi-Bred, 1989) provides a model of the type of patent claim that would be presented for a variety obtained in this way. The application was rejected because it contained no description of the method by which the variety had been obtained. Although seeds of the variety had been deposited with a culture collection, in conformity with the widely established practice for microorganisms, the court did not accept the deposit as a substitute for a written description of the method of production. The Canadian court is in this respect out of line with the courts of the USA, Europe and Japan. The claim read:

> A variety of soybean plant characterised by having the following characteristics:
> Seeds:
>
> | | |
> |---|---|
> | shape | oblong |
> | surface | sometimes wrinkled |
> | seed coat color | medium yellow |
> | seed coat luster | shiny |
> | hilum color | light gray |
> | weight | 18–20 grams per 100 seeds |
> | cotyledon color | yellow |
>
> and also, exhibiting longitudinal discoloration of the seed coat stemming from the hilum, visible in the event that the plant has experienced considerable environmental stress.
> Leaves:
>
> | | |
> |---|---|
> | color | medium green |
> | shape | ovate |
> | plant pubescence color | medium gray |
> | plant height | 27–35 inches |
> | plant type | with intermediate canopy, i.e. intermediate between slender and bushy |
> | plant habit | indeterminate |
>
> Pods:
>
> | | |
> |---|---|
> | color | brown |
> | set | scattered |
> | flower color | purple |
> | hypocotyl color | purple |
> | lodging score | 2–3, on a scale of 1–5 |
> | maturity group | 0 |
>
> said variety resembling the soybean variety Corsoy with respect to plant

shape, seedling pigmentation and leaf characteristics and the variety Portage with respect to seed size, and the variety Altona with respect to seed shape, and the variety Hardome with respect to color of hilum and is further characterized by being resistant to the fungus *Phytophthora megasperma* var. *sojae* (races 1 and 2).

This claim is based essentially on a listing of phenotypical properties. It might be difficult in many such cases to identify an inventive concept in any one such property or in a combination of such properties. This concrete example could help to clarify the issues in discussions between patent and UPOV circles which have hitherto often been at cross-purposes for want of a common understanding on terminology.

THE INTERPRETATION OF EPC ARTICLE 53(B)

A recent decision of the EPO Technical Appeal Board (Plant Genetic Systems, 1995) has upturned the hitherto prevailing interpretation of EPC Article 53(b). Plant Genetic System's European patent 242,236 was directed to transgenic plants containing in their cells a gene which conferred resistance to the herbicide Basta. The most important claim (claim 21) was to

> Plant, non-biologically transformed, which possesses, stably integrated into the genome of its cells, a foreign DNA nucleotide sequence encoding a protein having non-variety-specific enzymatic activity capable of neutralizing or inactivating a glutamine synthetase inhibitor under the control of a promoter recognised by the polymerase of said cells.

The patent also had claims to the methodology for transforming the plant, and claims to the vectors, plant cells and seed. It is important to note that the claims were not limited to particular plant species but referred to 'plants' in general. Until this patent was challenged the EPO had been willing to allow patents for plants defined in this generalized way, i.e. in non-variety-specific terms.

The patent was opposed by Greenpeace, who based their arguments on both limbs of Article 53 of the EPC, set out below.

> European patents shall not be granted in respect of:
> (a) inventions the publication or exploitation of which would be contrary to '*ordre public*' or morality, provided that the exploitation shall not be deemed to be so contrary merely because it is prohibited by law or regulation in some or all of the Contracting States;
> (b) plant or animal varieties or essentially biological processes for the production of plants or animals: this provision does not apply to microbiological processes or the products thereof.

The main attack on the patent was based on the morality and *ordre public* provisions of Article 53(a), the argument being that it was immoral to 'own' plants, which were the common heritage of mankind. Greenpeace supported this by producing results of surveys/opinion polls taken in Sweden (only farmers were consulted) and Switzerland.

The Technical Appeal Board considered the morality objection in depth and rejected it. The Board set out principles which they considered relevant to the assessment of such objections and their decision will be of greater use in cases where this objection is more appropriate than in one relating to plant biotechnology inventions. The Board considered the survey data as unrepresentative of attitudes in Member States. Indeed the Board evidently considered the morality objection misconceived in a case of this kind. As regards *ordre public* the Board would have considered this if there had been any evidence that exploitation of the patent would 'seriously prejudice the environment'. No such evidence was produced by Greenpeace.

But Greenpeace had also taken the Article 53(b) objection, arguing that the claims to plants and seeds would cover varieties formed from them and that essentially biological processes were involved. It was argued that the claims 'although cleverly drafted in general terms, were in reality meant to cover plant varieties', which would be contrary to Article 53(b). Furthermore, 'when a claim covered something which was unpatentable, the whole claim was bad'.

Greenpeace must have been surprised to find that whilst they had lost on 53(a), they were to win on 53(b). The Appeal Board was clearly influenced by the fact that in the specific patent examples of producing the transgenic plant, the process began with named varieties. The Board noted that claim 21 was not drafted in terms of a variety 'because there is no reference to a single botanical taxon of the lowest-known rank' but it held that the claim to transgenic plants 'includes within its scope known plant varieties which have been genetically modified so as to be herbicide-resistant ...' and was therefore not allowable under Article 53(b). The Board also said that the claim 'embraces' and 'encompasses' plant varieties, and it was therefore an attempt to evade the prohibition.

The Board also pointed to the new definition of a variety as given in the revised UPOV 1991 and held that the genetically modified plants were themselves new varieties according to the new definition. The Board held furthermore that the claim could not be allowed under the exception provided by the second half of Article 53(b) (the microbiological process exception) since the process of producing and propagating the transgenic plants, although it involved a microbiological step, was not a microbiological process when considered as a whole.

The Board allowed the claims to the transformation process and claims to plant cells but also rejected claims to plant cells when 'contained in a plant'. The research worker and others in industry must find this a strange result indeed. Why allow a process to be patented if the patent cannot also claim the novel product? This becomes even harder to answer when one realizes that, under EPC law, a process patent automatically gives protection for the direct product of the process.

Plant Genetic Systems appealed to the Enlarged Board of Appeal, which can review decisions of the Technical Boards in certain circumstances, including those where Technical Board decisions are inconsistent with one another. The Enlarged Board did not endorse the first part of the Technical Board's analysis (that the claim 'embraced' varieties). On their second point (that the transgenic plants were varieties) the Enlarged Board expressed no opinion, holding that it could not intervene because this was a new point which involved no inconsistency with previous decisions.

In view of this unsatisfactory decision of the Enlarged Board, the decision of the Technical Board therefore stands as authority which the EPO Examining Division now feel obliged to follow. Although the process technology can still be patented, the specific refusal of product claims to transgenic plants is a setback for European jurisprudence and for the plant biotechnology industry.

From the legal viewpoint, European patent attorneys will find it hard to justify this decision to their clients, for the following reasons.

1. The UPOV 1991 definition of 'variety' is much broader than any definition existing at the time the EPC was drawn up. It should therefore not be used to interpret the EPC exclusion.
2. UPOV 1991 was not in force at the time of the Technical Board's decision (and is not yet in force, at the time of writing).
3. The UPOV 1991 definition restricts a variety to the lowest known taxon whereas all the claims in this patent were directed to 'plants', presumably the highest known taxon.
4. Previous Technical Boards of Appeal have laid down the principle that all exclusions in the EPC must be construed as narrowly as possible (the total reverse of the reasoning in this case).

Another consideration is whether the decision would have been different if the specific patent examples had not mentioned named varieties. But plant genetic manipulation is now in use by most plant breeders, and is combined with their traditional methods. Seeds or plants sent to the market are usually varieties. Genetic engineering will surely be applied to achieve the further improvement of known varieties including those that represent the best that traditional breeding has given to agriculture.

The Appeal Board found itself bound by a legal provision which was never clear and which is now out of tune with the needs of the plant breeding industry. The invention in this patent is not one which could have been conveniently and effectively protected by plant variety rights and it deserved generic product patent protection. The anomalous decision in this case is welcomed only by those who resist the application of biotechnology to agriculture. Unless a future decision can redraw the proper line of demarcation between plant variety protection and patent protection so as to allow applicants to choose the type of protection which best serves their needs, there must be a good case for the abolition of EPC Article 53(b).

PATENTING PLANT GENETIC MATERIAL

In comparison with the problems of conflict between the legal systems for protecting plants, the patenting of plant genes in Europe is relatively straightforward and comparable to US patent practice. Genes are a special example of the broad class of naturally occurring materials which in appropriate circumstances can be patented. Where it is necessary to isolate and characterize a natural product and to devise a process for producing it, or using it, in quantity before it can be utilized by man for any practical purpose, the patent law offers scope for protection. Mere pre-existence of the substance, in admixture with vast quantities of other materials, is insufficient to contradict this view. This is the declared position of the WIPO (WIPO, 1988), of the EPO (EPO, 1995), and of the European Commission (EU Directive, 1995).

CONCLUSION

The use of biotechnology to modify the genetic constitution of plants is capable of producing inventions which ought to be accommodated within the framework of the patent law without serious difficulty. Patent practice is proceeding on this basis in countries with a generous patent law and patents for plants are being granted in appropriate instances. Nothing should be done to halt this trend. It is unrealistic to try to pre-empt the role of patents by seeking a dominant position for a new and improved UPOV system. No matter what improvements are made to plant variety law, the protection is unlikely ever to reach the level offered by patents because it inherently lacks generic character, being always pitched at the level of specific varieties.

Nevertheless, it would be equally undesirable for the patent system to interfere with the law of PVR working within its own proper sphere of operation. For certain types of plants the securing of these rights is closely bound up with obtaining national listing of the variety before its commercialization. This connection reinforces the necessity for a flourishing PVR system for which there is no easy substitute or alternative. However, so long as patent law excludes the granting of patents for plant varieties it is vital that the limits of this exclusion be clearly apparent in order to avoid confusion in the protection of those plant-related inventions which fulfil the criteria of patentability.

REFERENCES

Byrne, N. (1991) *Commentary on the Substantive Law of the 1991 UPOV Convention.* University of London Centre for Commercial Studies.
Ciba-Geigy (1984) Propagating material and lubrizol genetics/hybrid plants. *Official Journal of the European Patent Office* 3/1984, 112.
EPC (1973) 1995 Edition available from European Patent Office, Munich.
EPO (1995) *European Patent Office Guidelines*, Part C, Chapter IV, European Patent Office, Munich, 2.3, 36.
EU Directive (1988/1995) *European Commission – Proposal for a Directive on the Legal Protection of Biotechnological Inventions* – Comm (1988) 496 final SYN 159, Brussels, 17 October 1988, and *Revised Directive* of 13 December 1995, Comm (1995) 661 final, Brussels.
European Council (1994) *European Council Regulation No. 2100/94 on Community Plant Variety Rights*, Brussels.
Lubrizol (1990) Lubrizol Genetics. *Official Journal of the European Patent Office* 3/1990, 59–62.
Pioneer Hi-Bred (1989) 25 *Canadian Patent Reporter* (3d) 257.
Plant Genetic Systems (1995)(a) T365/96 Technical Board of Appeal decision February 1995. *Official Journal of the European Patent Office* 8/1995, 545. (b) Enlarged Board of Appeal decision G3/95, November 1995. *Official Journal of the European Patent Office* 4/1996, 169.
Straus, J. (1984) Patent protection for new varieties of plant produced by genetic engineering: should double protection be prohibited? *International Review of Industrial Property and Copyright Law* IIC, 426.
UPOV (1961) *International Convention for the Protection of New Plant Varieties.* World Intellectual Property Organization, Geneva.
UPOV (1978) *International Convention for the Protection of New Plant Varieties.* Additional Act of November 10 1972, and Revised text of October 23 1978, World Intellectual Property Organization, Geneva.
UPOV (1991) *Diplomatic Conference Document DC/91/138.* Revised 19 March 1991, UPOV publication no. 221 (CE), World Intellectual Property Organization, Geneva.
WIPO (1988) *Revised Suggested Solutions Concerning Industrial Property Protection of Biotechnological Inventions.* Documents BioT/CE/IV/3 and BioT/CE/IV/4. World Intellectual Property Organization, Geneva.
WIPO/UPOV (1990) *Report of the Committee of Experts on the Interface between Patent Protection and Plant Breeders' Rights.* Document WIPO/UPOV/CE/1/4. World Intellectual Property Organization, Geneva.

Patent Cooperation Treaty Members

As of 5 September 1997 all of the following countries are members of the Patent Cooperation Treaty (PCT):

Albania, Armenia, Australia, Austria, Azerbaijan, Barbados, Belarus, Belgium, Benin, Bosnia and Herzegovina, Brazil, Bulgaria, Burkina Faso, Cameroon, Canada, Central African Republic, Chad, China, Congo, Côte d'Ivoire, Cuba, Czech Republic, Democratic People's Republic of Korea, Denmark, Estonia, Finland, France, Gabon, Georgia, Germany, Ghana, Greece, Guinea, Hungary, Iceland, Indonesia, Ireland, Israel, Italy, Japan, Kazakstan, Kenya, Kyrgyzstan, Latvia, Lesotho, Liberia, Liechtenstein, Lithuania, Luxembourg, Madagascar, Malawi, Mali, Mauritania, Mexico, Monaco, Mongolia, Netherlands, New Zealand, Niger, Norway, Poland, Portugal, Republic of Korea, Republic of Moldova, Romania, Russian Federation, Saint Lucia, Senegal, Sierra Leone, Singapore, Slovakia, Slovenia, Spain, Sri Lanka, Sudan, Swaziland, Sweden, Switzerland, Tajikistan, Togo, Trinidad and Tobago, Turkey, Turkmenistan, Uganda, Ukraine, United Kingdom, United States of America, Uzbekistan, Vietnam, the former Yugoslav Republic of Macedonia, Yugoslavia, Zimbabwe.

Index